Long wave polar modes in semiconductor heterostructures

Long wave polar modes in semiconductor heterostructures

C. Trallero-Giner, R. Pérez-Alvarez and
F. García-Moliner

Pergamon

UK	Elsevier Science Ltd., The Boulevard, Langford Lane, Kidlington, Oxford OX5 1GB
USA	Elsevier Science Inc., 655 Avenue of the Americas, New York, NY 10010, USA
JAPAN	Elsevier Science Japan, 9-15 Higashi-Azabu 1-chome, Minato-ku, Tokyo 106, Japan

Copyright © 1998 Elsevier Science Ltd.

All rights Reserved, No part of this publication may be reproduced, stored in a retrieval system or transmitted in any form or by any means electronic, electrostatic, magnetic tape, mechanical, photocopying, recording or otherwise, without permission in writing from the copyright holders.

First edition 1998

Library of Congress Cataloging in Publication Data
A catalog record for this book is available from the Library of Congress

British Library Cataloguing in Publication Data
A catalog record for this book is available from the British Library

ISBN 0 08 0426948

Printed and bound in Great Britain by Redwood Books Ltd.

Contents

Preface   vii

List of symbols   xi

1. Phonons in bulk crystals   1
2. The long wave limit (bulk). Continuum approach   13
3. Polar optical modes in heterostructures   21
4. Surface Green Function Matching   51
5. Polar optical modes in layered structures   83
6. Quasi-1D semiconductor nanostructures   123
7. Quasi-0D semiconductor nanostructures   145

Index   163

Preface

This book is concerned with a phenomenological approach to the study of polar optical modes in semiconductor heterostructures. An accurate description of all details requires microscopic lattice dynamics calculations which, for these systems, entail rather heavy computation. The polar nature of these modes, with the long range Coulomb interaction, complicates the problem considerably and, in fact, rather than fully *ab initio* calculations, widespread resource to model making at some stage or other is often found. Lattice dynamics calculations still have the power to yield a detailed description — assuming the models involved to be reliable — but some of the details are actually not always exceedingly relevant, or even accessible to experiment and a simpler phenomenological approach may often be useful and desirable.

The establishment of a satisfactory phenomenological model, amenable to comparatively light computation and at the same time capable of giving a reliable account of polar optical modes in heterostructures has proved somewhat elusive for some time. Different proposals have been made which, while meeting with some partial degree of success, turn out to have some obvious limitations often amounting to severe formal inconsistencies. Yet, while these limitations are by now obvious and well known, resource to incomplete models is still often common practice. The issue concerns mainly — but not exclusively — the inability of many of these models to give a satisfactory account of both, mechanical and electrical matching boundary conditions simultaneously. The problem is really a very simple one of classical elasticity and electrostatics together with a careful analysis of the (matrix) differential operator involved, but an excessive hurry to obtain quick results has often prevented a relaxed analysis of the basic and very simple aspects.

A key issue, which is all too often ignored, is that when vector fields are matched at an interface the problem admits no purely longitudinal solutions. Yet most of these models start by assuming longitudinal optical modes for heterostructures. For given special conditions one can have solutions which in practice are nearly longitudinal to a good approximation, but as a general proposition an erroneous assumption which violates an existence theorem cannot provide a firm foundation for a formally correct model. Use of physical intuition and model making are of course common practice among physicists, but sometimes a minimum of rigour in the formal analysis is both, necessary and practical. In this case it is easy to see that the unwarranted assumption that the problem has longitudinal solutions leads to two alternatives which not only yield incompatible matching boundary conditions but also rely on different differential operators which can never yield amplitudes with the same spatial dependence. In a research field in which, partly due to the large potential for applications, there is a great deal of activity, it would seem desirable to have a phenomenological model which mostly experimentalists can use with ease and confidence to analyse experimental data without great mathematical or computational complication.

For the last few years we have been developing and using a long wave phenomenological model which, besides performing quite well in practice, is rid of inconsistencies and meets all the formal requirements without conflict. Having used this model in various contexts and for different types of heterostructures and after the suggestion of some colleagues with whom we have been in contact we decided to write this monograph where the phenomenological theory of long wave polar modes in heterostructures is discussed from this point of view. We pay attention to formal aspects, such as general analytical properties like Hermitean character of the differential matrix operator, completeness, formal rule for the normalisation of the field amplitude, etc. These formal considerations constitute the basis for an unambiguous derivation under general conditions — and within the frame of the phenomenological model — of a correct electron-phonon interaction Hamiltonian.

One of the purposes of the book is to provide useful tools for doing calculation to anyone who might be interested in practical applications. The mathematical properties of the basis functions for some characteristic situations are discussed in some detail and their use demonstrated in practical terms. We also present a Green function formalism as an alternative approach. Some examples of application show how this can provide a very useful way of doing practical calculations.

Piezoelectric waves are also — very briefly — commented as they have in common with polar optical modes the fact that they involve coupled mechanical and electrical vibrations, although the specific details are different. In both cases one may face the problem of matching coupled fields under conditions in which one of them is subject to special conditions at the surface or interface. The Green function analysis provides in this case a good instrument for the study of this problem. But the brief reference to piezoelecticity is only due to the intrinsic instructive value of the comparison between the two problems. The focus is on polar modes in semiconductor heterostructures. The symptoms are that the activity in this field will continue for some time and we hope that the material presented here may be useful to anyone interested in studying these problems without any other baggage than classical elasticity and electrostatics.

The state of the art is far more developed for planar heterostructures — quantum wells or barriers, superlattices, double barrier structures, etc — than it is for quantum wires and dots, but these are also the subject of substantial and growing activity and we also devote some space to discuss these. Although the main thrust of the book is in planar heterostructures, cylindrical or spherical geometries can also be studied with the same techniques and the purpose of the book is to present these, so anyone interested can use them. In a field in which the potential for device application is so large the activity is at times somewhat frantic and we thought it might be appropriate to try and introduce some systematics, setting on a practical basis a phenomenological model which can be utilised with relative ease in a way which gives a rather transparent physical picture while being basically sound. We hope that people interested in a phenomenological analysis which avoids heavy numerical computation might find this of some use.

Polar optical modes in semiconductors play an important role in many physical properties due to a significant electron-phonon interaction which often is the dominant scattering mechanism. Polaron effects and electronic, optical and electrooptical properties of various kinds can be affected by the inelastic scattering of electrons due to polar optical modes. The study of these phenomena is outside the scope of this book, which is concerned with the normal modes themselves; what are the eigensolutions for matched heterostructures involving one or more interfaces, how — with what techniques — can these be studied and what is the physical nature of the normal modes. Altogether the spectrum of these modes for heterostructures has significant distinctive features and the purpose of this book is to present the state of the art concerning the study of this problem. The link between this and the theory of many physical processes where these modes intervene is through a correct formulation of an electron-phonon interaction Hamiltonian and so we devote explicit attention to this problem and stop there.

Bibliographical references are given at the end of each chapter.

In the course of our work on these problems we have largely benefited from contacts and collaborations with some colleagues. It is a pleasure to acknowledge how much we owe to Víctor R. Velasco, Manuel Cardona, and Fernando Comas. We are also indebted to the Foundation of the Savings Bank of Castellón (Spain) for the help which enabled us to work together in the hospitable environment of the Jaume I University (Castellón) and in the Foundation's own premises. Both these institutions have in their statutes and explicit provisions an avowed comittment to peace. It is not by mere chance that one of the four sites of the UNESCO inspired European Peace University is physically located here and berthed by the two. In a world full of irrational conflicts and tensions it is a rare and soothing privilege to enjoy the facilities of two institutions where peace is given such high consideration and there write a book of physics which collects our common experience and the many valuable inputs received by many colleagues from different countries in spite of the doings of governments and politicians which do not always seems to aim at making a better world.

Havana and Castellón, June 1997.

List of symbols

a, b	Linear lattice period
a_0	Lattice constant in 3D
$\hat{b}$	Annihilation operator for phonons
$\hat{b}^\dagger$	Creation operator for phonons
c	Speed of light in vacuum
$\mathbf{D}$	Electric displacement vector
$\hat{\mathbf{D}}$	Differential operator
e	Electron charge
$\mathbf{e}$	Unit vector (e.g. of $\mathbf{u}$)
e	Base of natural exponential
$\mathbf{E}$	Electric field
$\mathbf{F}$	Tetrafield (3.56)
$\mathcal{F}$	Projection of $\mathbf{F}$ (4.31)
$\mathbf{G}$	Green function $\mathbf{G}(z, z')$
$\mathcal{G}$	Projection of the Green function. $\mathcal{G}(z) = \mathbf{G}(z, z)$
$'\mathbf{G}$	Derivative of $\mathbf{G}$ with respect to the first argument
$'\mathcal{G}^\pm$	Projections of derivatives of the Green function relevant to SGFM (Eq. (4.10))
H	Hamiltonian
$\mathcal{H}$	Hamiltonian density
$\hbar = h/2\pi$	Dirac constant
$k, \mathbf{k}$	Wavevector
k_L, k_T	See Eqs. (4.20) and (4.21)
$\mathbf{K}_m$	Primitive translation in the reciprocal space
L	Lagrangian
$\mathcal{L}$	Lagrangian density
$\hat{\mathbf{L}}$	Differential operator
$\hat{\ell}$	Angular momentum operator $(-i\mathbf{r} \times \nabla)$
m, M	Mass
N	Number of unit cells in the crystal
$\hat{n} = \hat{b}^\dagger \hat{b}$	Particle number operator for phonons

P	Polarisation
p_n	Momentum of atom n
Q_T, Q_L	See Eqs. (3.33) and (3.37)
q	Bloch wavevector associated with superperiodicity
r	Position vector (x, y, z)
R$_n$	Position vector of the unit cells in the crystal
t	Time
T	Kinetic energy
U	Potential energy
$\mathbf{u} = \mathbf{u}_+ - \mathbf{u}_-$	Relative atomic displacement of cation and anion in the unit cell *except* in Figure 5.4 and corresponding text.
u_{ij}	Strain tensor (2.10)
V	Volume
v_c	Unit cell volume
v_s	Speed of sound
α	Tensor coupling **u** and **E**
δ	Tensor coupling **E** with itself
$\beta_L\ (\beta_T)$	Parabolicity parameters of the longitudinal (transverse) bulk phonon branch
γ	Tensor coupling **u** with itself
ϵ	Dielectric constant tensor
$\epsilon(\omega)$	Frequency dependent dielectric constant
ϵ_0	Static dielectric constant
ϵ_∞	Optical dielectric constant
η	Small infinitesimal part of the eigenvalue
κ	Wavevector parallel to the interfaces (2D)
λ_{ijkl}	Fourth rank tensor associated with the internal 'stresses' of the medium
μ	Reduced mass
$\|0, 0, \ldots, \nu_k, \ldots, 0, 0>$	The state with ν_k phonons with wavevector k
π	Momentum density
ρ	Reduced mass density
ρ	Vector position in the (x, y) plane
ρ_P	Polarisation charge
σ	Number of atoms in the unit cell
σ	'Stress' tensor
φ	Electric potential
Ω	Eigenvalue (in general)

List of symbols

Ω_0	Normalization constant for eigenvectors (3.74)
ω	Frequency
$\omega_s(\mathbf{k})$	Frequency of the branch s at wavevector $\mathbf{k}$
ω_L	Frequency of the bulk longitudinal optical mode at the Γ point of the Brillouin Zone
ω_T	Frequency of the bulk transverse optical mode at the Γ point of the Brillouin Zone
Ψ	Auxiliar scalar function (3.30)
$\boldsymbol{\Gamma}$	Auxiliar vector function (3.30)

CHAPTER ONE

Phonons in bulk crystals

We start with a very brief reminder of the main basic concepts in the theory of lattice dynamics in bulk crystals. This subject is of course abundantly — and well — covered in numerous texts of various kinds [1–6]. Here we shall simply introduce in an organised manner the notation to be followed and the basic elements — often simply the symbols and notation — to be used later.

Since simple systems usually illustrate the features of real systems, we start with a brief summary of linear chains and after that we collect the phonon dispersion and parameters of the semiconductors we shall mostly deal with, i.e. GaAs and AlAs and their ternary compounds.

Linear lattices

Monoatomic linear chain

We begin with the case of a linear lattice with N equidistant atoms of mass M and period a, interacting with an harmonic potential U and a nearest neighbour force constant γ_0 and look only at longitudinal vibrations. Then the atomic displacements in the normal mode k are

$$\xi_n = A_k e^{i(kna-\omega t)} \tag{1.1}$$

with corresponding eigenvalue

$$\omega = 2\sqrt{\frac{\gamma_0}{M}} |\sin(ka/2)| . \tag{1.2}$$

Let us define the collective coordinates A_k

$$\xi_n = \frac{1}{\sqrt{N}} \sum_k A_k e^{ikna} \tag{1.3}$$

with inverse relation

$$A_k = \frac{1}{\sqrt{N}} \sum_k \xi_n e^{-ikna} . \tag{1.4}$$

The requirement ξ_n = real then implies that $A_k^* = A_{-k}$. In these coordinates the phonon Hamiltonian takes the form of N decoupled oscillators. For each oscillator we can introduce the operators $\hat{b}_k$ and $\hat{b}_k^\dagger$ and then

$$\hat{H} = \sum_k \hbar\omega(k)\left(\hat{b}_k^\dagger \hat{b}_k + \frac{1}{2}\right), \qquad (1.5)$$

$$\left[\hat{b}_k, \hat{b}_{k'}\right] = 0 \quad ; \quad \left[\hat{b}_k^\dagger, \hat{b}_{k'}^\dagger\right] = 0 \quad ; \quad \left[\hat{b}_k, \hat{b}_{k'}^\dagger\right] = \delta_{kk'}. \qquad (1.6)$$

If $|0, 0, \ldots, v_k, \ldots, 0, 0>$ is the state with v_k phonons in mode k, then

$$\begin{aligned}
\hat{b}_k |v_k> &= \sqrt{v_k}\ |v_k - 1>, \\
\hat{b}_k^\dagger |v_k> &= \sqrt{v_k + 1}\ |v_k + 1>, \\
\hat{n}_k &= \hat{b}_k^\dagger \hat{b}_k, \\
\hat{n}_k |v_k> &= v_k\ |v_k>, \\
E_0 &= \frac{1}{2}\sum_k \hbar\omega(k), \\
E &= E_0 + \sum_k v_k \hbar\omega(k).
\end{aligned} \qquad (1.7)$$

Diatomic linear chain

We recall the well-known fact that a linear monoatomic chain with period a can also be formally described as having a period $2a$, $3a$, etc. The dispersion relation curve $\omega(k)$ is then folded in a smaller Brillouin Zone (BZ) and the flatness property $\partial\omega/\partial k|_{BZ\ border} = 0$ no longer holds. If we modify the system in some way so that the fictitious period becomes real, then some gaps appear at the Brillouin Zone border of the initially folded representation and the flatness property reappears. With a view to the eventual treatment of the superlattices it is appropriate to look here in this way at the passage from the monoatomic to the diatomic chain, so we now consider the case of two different atoms (masses M_α, $\alpha = 1, 2$) in the unit cell interacting with nearest neighbour force constant γ_0. For simplicity we assume the atoms to be equidistant so that $b = 2a$ is now the real period. In terms of collective coordinates A_k we now have

$$\xi_{n,\alpha} = \frac{1}{\sqrt{NM_\alpha}} \sum_k A_k e_\alpha(k)\ e^{iknb},$$

$$e_\alpha(-k) = e_\alpha(k) \quad ; \quad A_k^* = A_{-k}, \tag{1.8}$$

where A_k is the amplitude of the wave and $e_\alpha(k)$ is the polarisation vector (one-component vector in this 1D case) for α atom.

Then the Lagrangian is given by [1]

$$L = \frac{1}{2}\sum_{k,\alpha} e_\alpha^2(k)\dot{A}_k\dot{A}_{-k} - \frac{1}{2}\sum_{k,\alpha,\beta} D_{\alpha\beta}(k)e_\alpha(k)e_\beta(k)A_k A_{-k} \tag{1.9}$$

with

$$D_{11} = 2\gamma_0/M_1 \quad ; \quad D_{12} = -\gamma_0/\sqrt{M_1 M_2}\left(1 + e^{ikb}\right)$$
$$D_{21} = D_{12}^* = D_{12}(-k) \quad ; \quad D_{22} = 2\gamma_0/M_2 \tag{1.10}$$

and the Lagrange equations are

$$e_\alpha(k)\ddot{A}_k + \sum_\beta D_{\alpha\beta}(k)e_\beta(k)A_k = 0 \quad ; \quad \alpha = 1, 2. \tag{1.11}$$

These and $\ddot{A}_k = -\omega^2 A_k$, yield the dispersion relations

$$\omega_s^2(k) = \gamma_0\left[\frac{1}{\mu} + (-1)^s\sqrt{\frac{1}{\mu^2} - \frac{4}{M_1 M_2}\sin^2\left(\frac{kb}{2}\right)}\right], \tag{1.12}$$

$$s = 1, 2,$$

where μ is the reduced mass given by

$$\mu = \frac{M_1 M_2}{M_1 + M_2}. \tag{1.13}$$

For $ka \ll 1$

$$\omega_1(k) \approx \frac{kb}{2}\sqrt{\frac{2\gamma_0}{M_1 + M_2}} = v_s k \quad ; \quad \omega_2(k) \approx \sqrt{\frac{2\gamma_0}{\mu}} \tag{1.14}$$

where v_s is the speed of sound. At the Brillouin Zone border

$$\omega_1(\pi/a) = \sqrt{\frac{2\gamma_0}{M_{max}}} \quad ; \quad \omega_2(\pi/a) = \sqrt{\frac{2\gamma_0}{M_{min}}}, \qquad (1.15)$$

with

$$M_{max} = \text{Max}\{M_1, M_2\} \quad ; \quad M_{min} = \text{Min}\{M_1, M_2\}, \qquad (1.16)$$

and the displacements of the atoms in the unit cell are related by

$$\left(\frac{\xi_{n2}}{\xi_{n1}}\right)_{k,s} = \frac{e_{2s}(k)}{e_{1s}(k)} \sqrt{\frac{M_2}{M_1}},$$

$$\frac{e_{2s}(k)}{e_{1s}(k)} = \frac{\frac{2\gamma_0}{M_1} - \omega_s^2(k)}{2\gamma_0/\sqrt{M_1 M_2}}. \qquad (1.17)$$

In the above equations ω_1 represents the *acoustical branch*, and ω_2 the *optical branch*.

In terms of phonon creation and annihilation operators

$$\hat{H} = \sum_{k,s} \hbar\omega_s(k) \left[\hat{b}^\dagger_{ks}\hat{b}_{ks} + \frac{1}{2}\right], \qquad (1.18)$$

$$\hat{\xi}^{(s)}_{n\alpha} = \sqrt{\frac{\hbar}{2N M_\alpha}} \sum_{k,s} \frac{e_{\alpha s}(k)}{\sqrt{\omega_s(k)}} \left(\hat{b}_{ks} + \hat{b}^\dagger_{ks}\right) e^{ikna}. \qquad (1.19)$$

Qualitatively similar results are obtained when we treat the case of two — different or equal — nonequidistant atoms in the unit cell.

In these systems we have $2N$ degrees of freedom and $2N$ second order coupled ordinary differential equations, N being the number of unit cells. The introduction of collective coordinates allows us to solve the problem with a system of $2N/N = 2$ algebraic equations.

In the acoustic branch ω tends to zero when $k \to 0$, and then in the long-wavelength limit

$$\frac{\xi_{n2}}{\xi_{n1}} \approx 1. \qquad (1.20)$$

On the other hand, for the optical branch and $k \approx 0$, $\omega \approx \sqrt{2\gamma_0/\mu}$ and

$$\frac{\xi_{n2}}{\xi_{n1}} \approx -\frac{M_1}{M_2}. \tag{1.21}$$

The standard use of the Born-von Karman conditions to count the number of states yields again a total number of modes equal to the number of atoms in the Born-von Karman ring, but now there are two branches in correspondence with the number of atoms in the unit cell and a gap between the two. If we take p cells as a fictitious period then the dispersion relation scheme is folded p times and we loose the flatness property in the neighbourhood of the Brillouin Zone border. Again, if the system is modified so the fictitious period becomes real, then the property is recovered and some gaps open at the Γ and X points of the Brillouin Zone.

General comments on linear lattices

We have examined simply longitudinal modes in linear chains of atoms with only nearest neighbour interactions, which suffices to set the scene and bear out the main qualitative features. For the diatomic chain we have assumed equidistant atoms with only a mass difference. Again, a re-examination of this problem with some changes — e.g. equal or different atoms but with a nearest neighbour spacing $a \neq b/2$ — does not change the key qualitative features and if we recover the monoatomic chain by making all atoms and/or interatomic distances equal, then we obtain one branch but artificially folded.

The consideration of p atoms of the same or different classes distributed in the unit cell interacting with an arbitrary number of neighbours implies quantitative changes in the dispersion relation curves and normal modes. But the general picture is still valid, i.e. the states are labelled by the 1D wavevector k spanning the first Brillouin Zone, the number of branches (in one dimension) agrees with the number of atoms in the unit cell, the number of states in a branch is determined by the number of unit cells in the crystal (Born-von Karman or periodic boundary condition), there is always an acoustic branch with frequency tending to zero in the long-wavelength limit, and there are $p-1$ optical branches in the upper side of the spectrum.

In particular if we consider the linear diatomic chain and we call γ_0 the force constant for atom 1–atom 2 interaction, γ_1 the force constant for atom 1–atom 1 interaction and γ_2 the force constant for atom 2–atom 2 interaction, then we arrive at the following final formulae for the frequency and its derivatives at Γ and X points of the Brillouin Zone:

$$\omega(\Gamma) = \sqrt{\frac{2\gamma_0}{\mu}} \quad ; \quad \omega(X) = \begin{cases} \sqrt{2(\gamma_0 + 2\gamma_1)/M_1} \\ \\ \sqrt{2(\gamma_0 + 2\gamma_2)/M_2} \end{cases}, \qquad (1.22)$$

$$\frac{\partial \omega(\Gamma)}{\partial(qd)} = \sqrt{\frac{\gamma_0/2 + \gamma_1 + \gamma_2}{M_1 + M_2}}, \qquad (1.23)$$

$$\frac{\partial^2 \omega(\Gamma)}{\partial(qd)^2} = -\frac{\mu}{\omega(\Gamma)}\left[\frac{\gamma_0}{2M_1 M_2} - \frac{\gamma_1}{M_1^2} - \frac{\gamma_2}{M_2^2}\right]. \qquad (1.24)$$

Similar formulae hold for longer range interaction.

These simple considerations on linear chains will be applicable to many situations with 3D systems when, due to symmetry, the problem is effectively reduced to that of a linear chain.

Lattices in 3D

Let $\mathbf{R}_n$ be the vectors of the direct lattice and $\mathbf{K}_m$ the vectors of the inverse one. $M_\alpha, \alpha = 1, 2, \ldots, \sigma$, where σ is the number of atoms in the unit cell, α labels the different atoms and M_α is the mass of atom α. Let $\boldsymbol{\xi}_{n\alpha}$ be the displacement with respect to the equilibrium position of the atom α in the cell n. Then, in the harmonic approximation the vibrations of this 3D lattice are characterised by the Lagrangian

$$L = \frac{1}{2}\sum_{n,\alpha} M_\alpha \dot{\xi}_{n\alpha}^2 - \frac{1}{2}\sum_{n,\alpha}\sum_{p,\beta}\sum_{i,j=1}^{3} U_{n\alpha,p\beta}^{ij} \xi_{n\alpha}^i \xi_{p\beta}^j,$$

$$U_{n\alpha,p\beta}^{ij} = \left(\frac{\partial^2 U}{\partial x_{n\alpha}^i \partial x_{p\beta}^j}\right)_{\mathbf{R}_{n\alpha},\mathbf{R}_{p\beta}}. \qquad (1.25)$$

U is the total potential energy of the interaction between the atoms; i and j label the Cartesian components; $\mathbf{R}_{n\alpha}, \mathbf{R}_{p\beta}$ are the equilibrium positions for $(n\alpha)$ and $(p\beta)$ atoms respectively. The parameters $U_{n\alpha,p\beta}^{ij}$ satisfy well-known symmetry relations. Transforming to collective coordinates A_k,

$$\boldsymbol{\xi}_{n\alpha} = \frac{1}{\sqrt{NM_\alpha}}\sum_k A_k \mathbf{e}_\alpha e^{i\mathbf{k}\cdot\mathbf{R}_n} \qquad (1.26)$$

with

$$e_\alpha(-\mathbf{k}) = e_\alpha(\mathbf{k}) \quad ; \quad A_{\mathbf{k}} = A^*_{-\mathbf{k}} \, , \qquad (1.27)$$

the Lagrangian and Hamiltonian adopt the form of uncoupled oscillators. For each $\mathbf{k} \in$ BZ we have a system of 3σ linear differential equations with constant coefficients. Then we look for the solutions in the form $A_{\mathbf{k}}(t) = A_{\mathbf{k}} e^{i\omega t}$ and we have

$$\sum_{\beta=1}^{\sigma}\sum_{j=1}^{3}\left\{\omega^2(\mathbf{k})\delta_{\beta\alpha}\delta_{ij} - \frac{U_{\beta\alpha}^{ji}(\mathbf{k})}{\sqrt{M_\alpha M_\beta}}\right\} A_{\mathbf{k}} e_\beta^j(\mathbf{k}) = 0 \, , \qquad (1.28)$$

$$\alpha = 1, 2, \ldots, \sigma \quad ; \quad i = 1, 2, 3..$$

The secular equation is then

$$\left\|\omega^2(\mathbf{k})\delta_{\beta\alpha}\delta_{ij} - \frac{U_{\beta\alpha}^{ji}(\mathbf{k})}{\sqrt{M_\alpha M_\beta}}\right\| = 0 \, . \qquad (1.29)$$

This equation has 3σ solutions for each $\mathbf{k}$, i.e. 3σ branches appear in the phonon spectrum: $\omega_s(\mathbf{k})$, $s = 1, 2, \ldots, 3\sigma$.

The frequency of three of these branches tends to zero as $\mathbf{k}$ tends to the Γ point of the Brillouin Zone. The other $(3\sigma - 3)$ branches have a nonzero limit for $\mathbf{k} \to \mathbf{0}$. For all the branches $\omega_s(\mathbf{k} + \mathbf{K}_n) = \omega_s(\mathbf{k})$. In cubic crystals the polarisation vector of one of the acoustic (optical) branches coincides in direction with $\mathbf{k}$; these branches are called *longitudinal*, while the rest of the branches are *transverse*. In other crystalline systems this property only occurs for symmetry points but it is customary to keep this nomenclature even in other directions.

In terms of creation and annihilation operators we have

$$\hat{\xi}_{n\alpha} = \sqrt{\frac{\hbar}{2NM_\alpha}} \sum_{\mathbf{k}s} \frac{e_{\alpha s}(\mathbf{k})}{\sqrt{\omega_s(\mathbf{k})}} \left(\hat{b}_{\mathbf{k}s} + \hat{b}^\dagger_{\mathbf{k}s}\right) e^{i\mathbf{k}\cdot\mathbf{R}_n} \, , \qquad (1.30)$$

$$\hat{H} = \sum_{\mathbf{k}s} \hbar\omega_s(\mathbf{k}) \left\{\hat{b}^\dagger_{\mathbf{k}s}\hat{b}_{\mathbf{k}s} + \frac{1}{2}\right\} \, . \qquad (1.31)$$

As in 1D lattices the Bloch theorem and Born von Karman conditions play a fundamental role in describing the normal modes. The arguments presented in section 1 concerning folding and band flatness hold equally in 3D, where a given direction is chosen and what vanishes is the normal derivative at the Brillouin Zone border. If one formally defines a fictitious period p times larger than the natural one in some direction, then the corresponding size of the Brillouin Zone is accordingly reduced and the phonon branches folded in this direction. The description would then look like being artificially anisotropic but it would be formally valid.

Phonon parameters for GaAs, AlAs and related ternary solid solutions

For the materials we shall mostly deal with there are numerous experimental and theoretical values of the parameters relevant to the study of phonon modes, often finding different estimates for the same parameter. Here we shall collect the most commonly accepted ones for GaAs, AlAs and their ternary compounds. All of these crystallise in the zincblende lattice and have two atoms per unit cell; then they all have the same Brillouin Zone and six phonon branches, three acoustic and three optical.

As usual we denote by ω_L (ω_T) the optical longitudinal (transverse) frequency at the centre of the Brillouin Zone. These branches are also characterised by the parabolicity parameters β_L and β_T in the neighbourhood of the Γ point, where a good approximation to the dispersion relation for long waves is

$$\omega_\ell^2 = \omega_L^2 - \beta_L^2 k^2 \; ; \quad \omega_t^2 = \omega_T^2 - \beta_T^2 k^2 \; . \tag{1.32}$$

In Table 1.1 we collect some general data and parameters describing the materials and their optical branches: a_0 (lattice constant), ρ (reduced mass density), ϵ_0 (static dielectric constant), ϵ_∞ (high-frequency dielectric constant), ω_L, ω_T, β_L^2, β_T^2, etc.

The values of a_0, ρ, ϵ_0, ϵ_∞, ω_L, and ω_T (these two frequencies at different points of the Brillouin Zone), are taken from the Tables of Landolt-Börnstein [7]. The rest of the parameters are estimated from the curves $\omega(k)$ and the effective linear chain along the (100) direction for the optical branches (see Eqs. (1.22)–(1.24)).

GaAs and AlAs can be mixed in any proportion and grown as a regular array of atoms. The lattice is then *fcc* as in pure materials. In $Al_xGa_{1-x}As$ the lattice constant varies linearly between the lattice constants of GaAs and AlAs. Some other properties can also be linearly interpolated by the empirical formulae [8, 9].

$$\rho(x) = 5.36 - 1.60\, x \; ; \; (\text{c.g.s.}) ,$$
$$\epsilon_0(x) = 13.18 - 3.12\, x \; ; \; (\text{e.s.u.}) ,$$

Table 1.1 Some phonon data of GaAs and AlAs. m_p is the proton mass and ρ is the reduced mass density. (A) means estimated value from the figures of Landolt-Börnstein [7]. (B) means analytical estimation from the effective linear chain with 2 neighbour interactions (see text).

		AlAs	GaAs
m_{cation} (m_p)		26.98	69.72
m_{anion} (m_p)		74.92	74.92
a_0 (Å)		5.660	5.65325
ρ (g/cm^3)		0.73	1.34
ϵ_0		10.06	13.18
ϵ_∞		8.16	10.89
$\omega_L(\Gamma)$(cm^{-1})		403.7	292.37
$\omega_T(\Gamma)$(cm^{-1})		361.7	268.50
$\omega_L(X)$(cm^{-1})		386.3	234
$\omega_T(X)$(cm^{-1})		391.3	254
β_L^2 (10^{-12})	(A)	0.77	2.91
	(B)	0.51	2.28
β_T^2 (10^{-12})	(A)	3.56	3.12
	(B)	1.00	3.57

$$\epsilon_\infty(x) = 10.89 - 2.73\,x\ ;\ \text{(e.s.u.)}\,. \tag{1.33}$$

However, this type of interpolation would not work for $\omega_L, \omega_T, \beta_L$ and β_T. It has been strongly argued on the basis of experimental evidence [10] that the ternary compound $Al_xGa_{1-x}As$ can be described in the *two-mode model*. We shall adopt this viewpoint. It then follows that if we study the matching to GaAs (AlAs) we must assign to the ternary alloy the values of the frequencies for the longitudinal and transverse modes found experimentally for the GaAs (AlAs)-like modes in this alloy [9]

$$\begin{aligned}\omega_{L,GaAs}(x) &= 292.37 - 52.83\,x + 14.44\,x^2\,; \\ \omega_{T,GaAs}(x) &= 268.50 - 5.16\,x - 9.36\,x^2\,. \end{aligned} \tag{1.34}$$

$$\begin{aligned}\omega_{L,AlAs}(x) &= 359.96 + 70.81\,x - 26.78\,x^2\,; \\ \omega_{T,AlAs}(x) &= 359.96 + 4.44\,x - 2.42\,x^2\,. \end{aligned} \tag{1.35}$$

Here and henceforth ω is always given in cm^{-1}, in accordance with usual practice in experimental papers.

The β_L and β_T parameters are not usually reported in the literature and they are not known for the different types of modes in alloys. We have estimated their values for the pure materials ($x = 0, x = 1$) from the experimental curves of Ref. [10].

We have also made the following assumption: For very low (high) concentrations of Al, that is for $x \approx 0$ ($x \approx 1$), we take dispersion laws with $\beta = 0$ for the AlAs (GaAs)-like modes. This assumption relies on the fact that for these situations the atoms in question are isolated and their phonon branches should be flat. For a given concentration x, we perform a linear interpolation between the values for $x = 0$ and $x = 1$. With ω in cm^{-1}, β is dimensionless and we obtain

$$\beta^2_{L,GaAs}(x) = 2.91 (1-x) 10^{-12},$$
$$\beta^2_{T,GaAs}(x) = 3.12 (1-x) 10^{-12} \quad (1.36)$$

$$\beta^2_{L,AlAs}(x) = 0.77 \, x \, 10^{-12},$$
$$\beta^2_{T,AlAs}(x) = 3.56 \, x \, 10^{-12} \quad (1.37)$$

This interpolation is admittedly gratuitous, but in fact we tested that the values of these parameters are altogether rather unimportant.

It is important to notice that all the input parameters involved have some dispersion, depending on the source, due in some cases to different sample preparation (ω_L, ω_T, etc). While the values used by different authors for these parameters are in fair agreement, this is not so for β_L and β_T — which, we stress, are dimensionless if ω is measured in cm^{-1}. One finds in the literature values of β_L^2 ranging from 2×10^{-12} [11] to 6.3×10^{-12} [12, 13] and values of β_T^2 ranging form 0 [11] to 3.12×10^{-12} [14]. It is also seen from Table 1.1 that theoretical estimations give different values.

References

1. C. Kittel, *Quantum Theory of Solids*, John Wiley & Sons, Inc. (1963).

2. N.W. Ashcroft and N.D. Mermin, *Solid State Physics*, Holt, Rinehart and Winston (1976).

3. A.O.E. Animalu, *Intermediate quantum theory of crystalline solids*, Prentice Hall, Inc. (1977).

4. B. Di Bartolo and R.C. Powell, *Phonons and resonances in solids*, John Wiley & Sons, Inc. (1976).

5. M. Born and K. Huang, *Dynamical Theory of Crystal Lattices*, Clarendon Press, Oxford (1988).

6. G.P. Srivastava, *The Physics of Phonons*, Adam Hilger, Bristol (1990).

7. Landolt-Börnstein, *Numerical Data and Functional Relationships in Science and Technology* Vol. III/17, Springer Verlag, Berlin (1982).

8. B. Jusserand and M. Cardona, in *Light Scattering in Solids V*, edited by M. Cardona and G. Güntherodt, Springer-Verlag, Heidelberg (1989).

9. S. Adachi, J. Appl. Phys. **58**, R1 (1985).

10. Z.P. Wang, D.S. Jiang and K. Ploog, Solid State Commun. **65**, 661 (1988).

11. N.C. Constantinou, O. Al-Dossary and B.K. Ridley, Solid State Commun. **86**, 191 (1993).

12. C. Guillemot and F. Clerot, Superlattices and Microstructures **8**, 263 (1990).

13. M. Babiker, J. Phys. C **19**, 683 (1986).

14. R. Pérez-Alvarez, F. García-Moliner, V.R. Velasco and C. Trallero-Giner, J. Phys.: Condensed Matter **5**, 5389 (1993).

CHAPTER TWO

The long wave limit (bulk). Continuum approach

Most of the experimental data convey really long wave information, so substantial effort has been devoted to setting up and using long wave phenomenological models which treat the medium as a continuum. For acoustic modes this can be obtained as the long wave limit of a discrete lattice dynamics model, but this is not so for the optical modes — which are the concern of this book — where contiguous atoms vibrate out of phase and so a different starting basis is needed to set up a long wave model. We discuss this problem in this chapter and derive a standard phenomenological model for polar optical modes from a Lagrangian formalism which meets all the requirements one might reasonably require. A fairly sound phenomenological model is thus obtained which provides a starting basis for the study of the matching problems discussed in the following chapters.

Acoustic and optical modes in one dimension

In 1D the long wave limit for the acoustic branch is easily obtained by noting that the atoms vibrate in phase. Then the atomic displacements of contiguous atoms are very close and their difference $\xi_{n+1} - \xi_n$ can be approximated as a power series expansion of a certain continuous function $\xi(z)$, i.e.

$$\xi_{n+1} - \xi_n \approx \xi'(z)\, a + \xi''(z) a^2/2, \tag{2.1}$$

where a is the period and z can be taken as $\xi_n + a/2$. Then the well-known equation of elasticity theory is obtained [1, 2]

$$\frac{\partial^2 \xi(z,t)}{\partial t^2} = v_s^2 \frac{\partial^2 \xi(z,t)}{\partial z^2}, \tag{2.2}$$

where v_s is the speed of sound in the linear chain (see Eq. (1.14)).

However, the long wave limit for optical modes cannot really be obtained in the same way, as the vibrations of contiguous atoms are out of phase and their amplitudes are very different — compare (1.20) and (1.21) — so a power series expansion like (2.1) is not possible and a different formulation is required.

Some attempts have been made to develop an envelope-function theory of optical phonons for inhomogeneous systems which has the simplicity of the effective-mass theory for electrons (see Ref. [3] and References therein). An elegant approach is to derive the envelope-function model directly from microscopic lattice dynamics [4]. We adopt a purely phenomenological approach, in which a set of equations of motion and boundary conditions is introduced. This theory is obtained by adding spatial dispersion terms to the dispersionless *Born-Huang equations* [2].

Optical modes in three dimensions

In order to introduce a long wavelength equation for optical modes we adopt a phenomenological approach taking into account the main physical facts, i.e. the existence of a polarisation on the scale of the unit cell and its macroscopic counterpart **P** coupling the mechanical oscillations $\mathbf{u} = \mathbf{u}_+ - \mathbf{u}_-$ with a macroscopic electric field **E**, $\mathbf{u}_+$ ($\mathbf{u}_-$) being the cation (anion) vibration amplitude. We initially postulate a linear constitutive relation of the form [2]

$$\mathbf{P} = \Gamma_{12}\mathbf{u} + \Gamma_{22}\mathbf{E} \ . \tag{2.3}$$

At optical frequencies **u** is negligible and $\mathbf{P} = (\epsilon_\infty - 1)/4\pi \mathbf{E}$, hence $\Gamma_{22} = (\epsilon_\infty - 1)/4\pi$. ϵ_∞ is the high-frequency dielectric constant. At low frequencies the two terms must be considered.

The total potential energy is given by

$$U = \rho \mathbf{u}^2 - \Gamma_{12}\mathbf{u} \cdot \mathbf{E} - \frac{1}{2}\Gamma_{22}\mathbf{E}^2 \tag{2.4}$$

and the corresponding equations of motion are

$$\rho \frac{\partial^2 \mathbf{u}}{\partial t^2} = -\rho \omega_T^2 \mathbf{u} + \Gamma_{12}\mathbf{E} \ , \tag{2.5}$$

$$\nabla \cdot \mathbf{D} = \nabla \cdot (\epsilon_\infty \mathbf{E} + 4\pi \Gamma_{12}\mathbf{u}) = 0 \ . \tag{2.6}$$

At $\omega \approx 0$, **u** is practically given by

The long wave limit (bulk). Continuum approach

$$\mathbf{u} = \frac{\Gamma_{12}}{\rho \omega_T^2} \mathbf{E} .$$

Then

$$\Gamma_{12}^2 = \rho \omega_T^2 \frac{\epsilon_0 - \epsilon_\infty}{4\pi} ,$$

ϵ_0 being the static dielectric constant. In the following we shall denote the constant Γ_{12}^2 as α^2.

In this approach, and in the quasistatic limit ($c \to \infty$) the electrostatic potential φ is related to the electric field in the standard way $\mathbf{E} = -\nabla \varphi$. Other physical parameters of the medium are: $\rho = (\bar{M}/v_c)$, the reduced mass density, $\bar{M}$, the reduced mass of the two ions and v_c, the unit cell volume.

It is not difficult to see that Eq. (2.5) leads to transverse nondispersive oscillations with frequency ω_T and longitudinal nondispersive oscillations with frequency ω_L related by the *Lyddane-Sachs-Teller relation*

$$\omega_L^2 = \omega_T^2 \frac{\epsilon_0}{\epsilon_\infty} . \tag{2.7}$$

So far this macroscopic phenomenological approach excludes dispersion. This is, in fact, the standard *Born-Huang* model. Now, in the same spirit we can introduce dispersion by adding an extra force term in Eq. (2.5). Then

$$\rho \frac{\partial^2 \mathbf{u}}{\partial t^2} = -\rho \omega_T^2 \mathbf{u} - \alpha \nabla \varphi - \nabla \cdot \boldsymbol{\sigma} , \tag{2.8}$$

where

$$\sigma_{ij} = -\rho(\beta_L^2 - 2\beta_T^2)(\nabla \cdot \mathbf{u})\delta_{ij} - 2\rho \beta_T^2 u_{ij} \tag{2.9}$$

can be considered as some kind of 'stress' tensor for isotropic media and

$$u_{ij} = \frac{1}{2}(\nabla_j u_i + \nabla_i u_j) \tag{2.10}$$

is the corresponding strain tensor.

The phenomenological parameters β_L and β_T can be obtained by a fitting to the phonon dispersion relations for the bulk case and are very close to the sound velocities in the medium (see section 1).

We emphasise that $\boldsymbol{\sigma}$ is not the usual stress tensor because $\mathbf{u}$ represents the relative displacement of the ions. However, the tensor $\boldsymbol{\sigma}$ displays all the usual symmetry properties of the stiffness tensor for an isotropic medium. A more general analysis considering the anisotropy of the media can be found in next Chapter. In what follows we shall assume :

$$\mathbf{u}(\mathbf{r}, t) = \mathbf{u}(\mathbf{r})e^{-i\omega t} \quad , \quad \varphi(\mathbf{r}, t) = \varphi(\mathbf{r})e^{-i\omega t} \quad . \tag{2.11}$$

Then equation (2.8) is cast in the form:

$$\rho(\omega^2 - \omega_T^2)\mathbf{u} = \alpha\nabla\varphi + \rho\beta_T^2 \nabla \times \nabla \times \mathbf{u} - \rho\beta_L^2 \nabla\nabla \cdot \mathbf{u} \quad ; \tag{2.12}$$

while equation (2.6) is written as :

$$\nabla^2 \varphi = \frac{4\pi\alpha}{\epsilon_\infty} \nabla \cdot \mathbf{u} \quad . \tag{2.13}$$

We summarise the main aspects of the model. We have a mechanical equation of motion for the polar optical mode vibrational amplitude $\mathbf{u}$ which is of the form

$$\rho(\omega^2 - \omega_T^2)\mathbf{u} + \nabla \cdot \boldsymbol{\sigma} - \alpha\nabla\varphi = 0, \tag{2.14}$$

or equivalently (2.12).

The harmonic oscillator part is contained in the first term. The second one has the nature of a dispersive mechanical term and for an isotropic medium is of the form (2.9). Simultaneously we have a Poisson equation for φ. The meaning of this is that φ is created by the polarisation charge $\nabla \cdot \mathbf{P}$ of the polarisation field

$$\mathbf{P} = \alpha\mathbf{u} + \frac{\epsilon_\infty - 1}{4\pi}\mathbf{E} = \alpha\mathbf{u} - \frac{\epsilon_\infty - 1}{4\pi}\nabla\varphi \tag{2.15}$$

associated to the polar optical vibration $\mathbf{u}$.

The situation is similar to that of the theory of piezoelectric waves in the sense that we have a coupling between the mechanical and electrical amplitudes ($\mathbf{u}$ and φ) but the details and the constitutive relations are different [2].

For the longitudinal modes it can be seen that Eqs. (2.13) and (2.14) are equivalent and we can work with either a dielectric model (2.13) or a mechanical equation of motion (2.14).

For a bulk homogeneous medium these can be decoupled and this yields at once the longitudinal (L) and transverse (T) modes with dispersion relations

$$\omega_\ell^2(\mathbf{k}) = \omega_L^2 - \beta_L^2 k^2 \quad ; \quad \omega_t^2(\mathbf{k}) = \omega_T^2 - \beta_T^2 k^2, \tag{2.16}$$

where $\mathbf{k}$ is the 3D wavevector.

A remark is here in order concerning the long wave approximation. As in the ordinary Debye model for acoustic modes, when the medium is described as a continuum the count of the number of degrees of freedom — hence the correct number of independent modes — is lost and one must artificially introduce a cutoff so the correct number of modes is restored. The dispersion relations (2.16) must be viewed in this light. Firstly, they constitute a good approximation to the phonon dispersion relations, known either from experimental information or from lattice dynamics calculations, only for small values of k. In practice these values often cover the experimentally interesting range of long waves. Secondly, the cutoff is also needed here. This has implications on the study of confined systems, e.g. a quantum well, which will be discussed in Chapter 5.

Fröhlich electron-phonon interaction

For later reference in the case of heterostructures (section 3) we summarise here the derivation of the electron-phonon (Fröhlich) interaction Hamiltonian. The model follows the line of *Born-Huang-like equations* for the classical vibrational field. We examine a semiconductor with two atoms per unit cell where the relative displacement is given by the vector $\mathbf{u}$ and the equation of motion becomes [2]

$$\rho \frac{\partial^2 \mathbf{u}}{\partial t^2} = -\rho \omega_T^2 \mathbf{u} - \alpha \nabla \varphi \ . \tag{2.17}$$

For a charge-free system Eqs. (2.15) and (2.17) are supplemented by the Maxwell equation

$$\nabla \cdot \mathbf{D} = 0 \ , \tag{2.18}$$

where

$$\mathbf{D} = \mathbf{E} + 4\pi \mathbf{P} \ . \tag{2.19}$$

Equations (2.17) - (2.19) are satisfied by a displacement field decomposed in the form $\mathbf{u} = \mathbf{u}_L + \mathbf{u}_T$, where $\mathbf{u}_L$ is the longitudinal displacement ($\nabla \cdot \mathbf{u}_L \neq 0$, $\nabla \times \mathbf{u}_L = \mathbf{0}$) and $\mathbf{u}_T$ the transverse part ($\nabla \cdot \mathbf{u}_T = 0$, $\nabla \times \mathbf{u}_T \neq \mathbf{0}$). The longitudinal displacement $\mathbf{u}_L$ is responsible for the electron-phonon interaction and Eq. (2.18) is then reduced to

$$\nabla^2 \varphi = 4\pi \nabla \cdot \mathbf{P} \tag{2.20}$$

with the boundary condition

$$|\varphi| < \infty \quad , \quad \forall \ \mathbf{r} \ . \tag{2.21}$$

The electron-phonon interaction is equal to $H_F = e\varphi$. Hence the Fröhlich Hamiltonian can be derived by solving Eq. (2.20). Thus

$$H_F = \int \frac{e\left[-\nabla \cdot \mathbf{P}(\mathbf{r}')\right]}{|\mathbf{r}-\mathbf{r}'|} dV' \ . \tag{2.22}$$

For the longitudinal electric field $\mathbf{E}_L$ and polarisation $\mathbf{P}_L$ we obtain

$$\mathbf{E}_L = -\left[\frac{4\pi\omega_L^2}{\epsilon^*}\right]^{1/2} \mathbf{u}_L \tag{2.23}$$

and

$$\mathbf{P}_L = \left[\frac{\omega_L^3}{4\pi\epsilon^*}\right] \mathbf{u}_L \ , \tag{2.24}$$

where $1/\epsilon^* = 1/\epsilon_\infty - 1/\epsilon_0$. The general expression for purely longitudinal vibration field is given by

$$\mathbf{u}_L(\mathbf{r}) = \sum_\mathbf{k} \mathbf{k} \left(A_\mathbf{k} e^{i\mathbf{k}\cdot\mathbf{r}} + A_\mathbf{k}^* e^{-i\mathbf{k}\cdot\mathbf{r}} \right) \tag{2.25}$$

with the condition $A_\mathbf{k}^* = -A_{-\mathbf{k}}$ which ensures that vector $\mathbf{u}_L$ is real. Note that $\mathbf{u}_L$ here is not an eigenvector, but the eigenvector expansion of the longitudinal vibration field and its amplitude is determined by the choice of the coefficients $A_\mathbf{k}$, which is not arbitrary. The $A_\mathbf{k}$ must in fact be chosen so that the quantised field $\mathbf{P}_L$ satisfies the commutation relation [5]

$$\left[\hat{\mathbf{P}}_L(\mathbf{r}), \hat{\pi}_L(\mathbf{r}')\right] = i\pi\delta(\mathbf{r}-\mathbf{r}') \ , \tag{2.26}$$

where $\hat{\pi}(\mathbf{r})$ is the canonical conjugate momentum equal to

$$\hat{\pi}_L(\mathbf{r},t) = \frac{4\pi\epsilon^*}{\omega_L^2} \frac{\partial \hat{\mathbf{P}}_L(\mathbf{r},t)}{\partial t} \ . \tag{2.27}$$

According to this, the polarisation operator has the form

$$\hat{\mathbf{P}}_L(\mathbf{r}) = \sum_\mathbf{k} \left(\frac{\hbar\omega_L}{8\pi\epsilon^* V}\right)^{1/2} \frac{\mathbf{k}}{|\mathbf{k}|} \left(e^{-i\mathbf{k}\cdot\mathbf{r}} \hat{b}_\mathbf{k} + H.C.\right) \ , \tag{2.28}$$

where $\hat{b}_\mathbf{k}, \hat{b}_\mathbf{k}^\dagger$ are the annihilation and creation operators. These second-quantisation operators satisfy the commutation relations (1.6). For further reference (section 3) we note that (2.25) and (2.28) are normal mode expansions of the corresponding fields, written down for $t = 0$. This involves implicitly the question of some suitably normalised eigenvectors, which need not appear explicitly in this analysis, but it will in section 3.

Substituting Eq. (2.28) into (2.22) and using the standard formula

$$\int e^{i\mathbf{k}\cdot\mathbf{r}} \frac{(\mathbf{r}' - \mathbf{r})}{|\mathbf{r} - \mathbf{r}'|^3} dV' = -4\pi i \frac{\mathbf{k}}{k^2} e^{i\mathbf{k}\cdot\mathbf{r}} \qquad (2.29)$$

we obtain the Fröhlich Hamiltonian for non-dispersive longitudinal optical phonons

$$\hat{H}_F(\mathbf{r}) = \sum_{\mathbf{k}} \left(C_{\mathbf{k}} e^{i\mathbf{k}\cdot\mathbf{r}} \hat{b}_{\mathbf{k}} + H.C. \right), \qquad (2.30)$$

$$C_{\mathbf{k}} = -ie \left(\frac{2\pi \hbar \omega_L}{\epsilon^* V} \right) \frac{1}{k}. \qquad (2.31)$$

References

1. L.D. Landau and E.M. Lifshitz, *Course of Theoretical Physics*, Vol. 7, 'Theory of Elasticity', Pergamon Press, Oxford (1970).

2. M. Born and K. Huang, *Dynamical Theory of Crystal Lattices*, Clarendon Press, Oxford (1988).

3. B.A. Foreman, Phys. Rev. **B52**, 12260 (1995).

4. H. Akera and T. Ando, Phys. Rev. **B39**, 6025 (1989).

5. H. Haken, *Quantum field Theory of Solids*, North Holland, Amsterdam (1977).

CHAPTER THREE

Polar optical modes in heterostructures

Having set up a phenomenological model in Chapter 2 the problem is now to match — at one or more interfaces — different media described by the same standard model. This is discussed in this chapter, where generally valid matching conditions are seen to follow formally from the mathematical form of the starting field equations once these have been accepted. A general important point emerging in the discussion is that in a matching problem involving a vector field — in our case the vibrational amplitude **u** — the matching conditions in general mix up the longitudinal and transverse components. There are exceptions to this, due to symmetry and geometry. For instance, in an isotropic medium — or in a cubic medium with a suitable geometry chosen to exploit high geometry directions — a shear mode propagating parallel to the surface is decoupled from the rest. In the standard theory of elastic surface waves this is the well-known *shear horizontal mode* and this is also present in the case of polar optical modes. But our concern is with the rest of the vibrational amplitude which besides being coupled to the electrostatic polarisation wave is also a saggittal mode, with longitudinal and transverse parts. We discuss the consequences of ignoring these facts and their relationship to some difficulties encountered by simplified models.

For most problems it suffices in practice to use an isotropic model and, on this basis, we present a general method of solution of the system of coupled differential equations where the standard phenomenological model is embodied. The effects of crystal anisotropy may be in practice comparatively small for some of the properties of interest of the heterostructures discussed in this book, but such effects should show up in some suitably arranged experiments, so we discuss briefly a general way of incorporating the effects of crystal anisotropy treated as a perturbation.

General canonical questions are also discussed which provide a formal basis for the unambiguous derivation of a correctly normalised electrostatic potential field from which an electron-phonon Hamiltonian can be obtained for any arbitrary heterostructure.

Introduction

This book focuses on low dimensional systems where the spatial confinement is responsible for many new physical properties. These effects have practical importance in the fabrication of new devices in micro and optoelectronics [1]. The high electron mobility in two-dimensional systems [2], the quantum well laser [3], the optical modulator for integrated optics [4], etc. are just a few examples of the technical applications of these novel systems. Besides these, the reduced dimensionality presents some basic or even fundamental aspects of physical interest as in optical properties, quantum Hall effect, transport, materials science and others. The lattice oscillations play an important role, especially for the description or studies of polaronic effects, scattering rates, free-carrier absorption of light, Raman scattering, etc. One of the most important factors from the theoretical point of view is the strength of electron-phonon interactions having great influence on the above mentioned physical magnitudes and device applications. Moreover, the phonon spectrum in low-dimensional heterostructures presents a great variety of new physical phenomena, such as optical confined and interface modes, acoustic folded modes, and surface modes. A review of phonons in low-dimensional systems has been given in Refs. [5, 6].

Since the pioneer work of Esaki and Tsu in semiconductor superlattices [7] and owing to the crystal growth contemporary technologies, fabrication of structures with quasi-2D, quasi-1D and quasi-0D spatial confinement are available with high quality. Good quality quantum well (QW) heterostructures, multiple quantum-wells, superlattices (SL), quantum well wires (QWW) and quantum dots (QD), grown along specific crystallographic planes or directions [8] have been reported. These semiconductor nanostructures are usually fabricated from weakly ionic materials such as III-V or II-VI compounds and their ternary alloys. In these materials, polar optical modes play an important role in many physical processes, as mentioned above.

The usual bulk Fröhlich Hamiltonian for the electron-longitudinal optical-phonon coupling has often been utilised to calculate several physical quantities for heterostructures (see [9] and References therein), but around the eighties experimental evidence confirmed that there are substantial differences in the electron-phonon interaction of semiconductor heterostructures if compared with the bulk case. Experiments of magnetoabsorption and cyclotron resonance in inversion layers [10], heterostructures [11] and superlattices [12] evidenced the difference in the electron-optical phonon interaction of the low dimensional structures.

Raman scattering measurements in multilayer heterostructures produced concrete evidence on the differences detected in the optical phonon spectrum. Resonant Raman scattering supplies information which is very useful in understanding phonon symmetry and the electron-phonon interaction mechanism.

The optical confined modes and interface modes have been clearly observed by Raman scattering experiments in several 2D systems [13-15]. Micro-Raman-spectroscopy permits to study the interface phonon dispersion and selection rules for confined optical modes in superlattices and multiple quantum wells [16, 17].

For acoustic phonons the spectrum is not strongly modified, as is in the case of polar optical phonons, since in the bulk acoustic frequency range of the constituent materials propagation along the superlattice growth direction is allowed and the effect of the superperiodicity is essentially a folding of the acoustic branch (see Chapter 1). For polar optical oscillations in layered structures the normal modes are rather confined in different layers and the penetration into the adjacent layer is negligible. This is a distinctive feature of the optical frequency region where the bulk phonon dispersion branches of the constituent semiconductors do not overlap. As a typical example we can mention the GaAs and AlAs optical phonon dispersion.

When the two optical branches overlap, then the modes are quasi-confined or resonant e.g. Si/Ge and InAs/GaSb. [18–21]. For these systems the microscopic calculations show that a small relative displacement amplitude is present in the adjacent layers. [18].

Besides the confined and quasi-confined phonons there appear new modes with frequencies between the bulk-like transverse optical and longitudinal optical frequencies of the constituent materials and their vibrational amplitude has a maximum at interfaces with an exponential decay into both adjacent layers. These are called interface modes and exhibit strong dispersion.

To describe the polar optical oscillations in semiconductor nanostructures of the types of quantum well, quantum well wire, quantum dot and superlattice, different theoretical approaches have been used. Generally, microscopic calculations lead to results in reasonable agreement with experiments, particularly in both quantum well and superlattice. [22–29]. Microscopic calculations of the phonon dispersion for GaAs quantum well wire embedded in AlAs have been also reported in Refs. [30–32]. The anisotropic character of the phonon dispersion in that kind of structure was analysed in [30]. Confined and interface phonons are studied in [32].

Models based on phenomenological continuum approaches have also been proposed for the investigation of polar optical oscillations of the semiconductor nanostructures for specified geometries. Oscillations for the long-wavelength limit can be studied within the framework of continuum treatments such as dielectric continuum model [33–39] or hydrodynamic model [37, 40–44]. The application of the above models to nanostructures led to disagreement with both microscopic calculations and first order resonant Raman measurements [22–27, 45–48]. Since experimental information on polar optical modes in semiconductor heterostructures concerns essentially the longwave limit, it is obviously desirable to have a simple tractable continuum model which with comparatively light computational requirements is capable of giving a satisfactory account of all the essential features established by both experimental evidence and more elaborate microscopic calculations. In the following we shall formally establish the basis of such a model from general considerations

Basic equations

An elementary presentation of a continuum model for optical modes in bulk media has been introduced in section 2.

In this section, we set up a long wavelength macroscopic model for polar optical phonons in nanostructures of arbitrary geometry. In order to study heterostructures, where the phenomenological parameters vary between contiguous domains, we now discuss a more formal derivation of the basic equations for a continuum model starting from an assumed form of a Lagrangian density $\mathcal{L}$. In the spirit of the classical theory of macroscopic media we introduce the nonrelativistic Lagrangian density $\mathcal{L}(\mathbf{u}; \frac{\partial \mathbf{u}}{\partial t}; \nabla\mathbf{u}; \nabla\varphi)$

$$\mathcal{L} = \frac{1}{2}\rho \left(\frac{\partial \mathbf{u}}{\partial t}\right) \cdot \left(\frac{\partial \mathbf{u}}{\partial t}\right) + \frac{1}{2}\mathbf{u} \cdot \boldsymbol{\gamma} \cdot \mathbf{u} - \mathbf{u} \cdot \boldsymbol{\alpha} \cdot \nabla\varphi + \frac{1}{2}(\nabla\varphi) \cdot \boldsymbol{\delta} \cdot (\nabla\varphi) + \frac{1}{8\pi}(\nabla\varphi) \cdot (\nabla\varphi) + \frac{1}{2}(\nabla\mathbf{u}) : \boldsymbol{\lambda} : (\nabla\mathbf{u}) \, . \quad (3.1)$$

This depends on the generalised coordinates $\mathbf{u}$, φ and involves just $\nabla\varphi$ as is required by gauge invariance. It satisfies the following requirements: (a) $\mathbf{u}$ and $\nabla\varphi$ appear quadratically leading to linear equations of motion; (b) dispersive oscillations are included up to quadratic terms in the phonon wavevector (c) electrostrictive effects (coupling of the electric field $\mathbf{E}$ with strain) are neglected.

In Eq. (3.1) $\boldsymbol{\alpha}, \boldsymbol{\delta}, \boldsymbol{\gamma}$ are second rank tensors; $\boldsymbol{\alpha}$ describes the coupling between mechanical vibration vector and the electric field, $\boldsymbol{\delta}$ couples the electric field with itself and $\boldsymbol{\gamma}$ describes the coupling of mechanical amplitude with itself and is related to natural oscillation frequencies of the medium. The tensors $\boldsymbol{\gamma}$, $\boldsymbol{\delta}$ are symmetric by their very definitions while $\boldsymbol{\lambda}$ is a fourth rank tensor which has the nature of an 'elastic stiffness' tensor and is associated with the energy density due to the internal stresses of the medium. This term leads to dispersive oscillations up to quadratic terms in the phonon wavevector. We define formally the 'stress' tensor $\boldsymbol{\sigma}$ as:

$$\boldsymbol{\sigma} = \boldsymbol{\lambda} : \nabla\mathbf{u} \, . \quad (3.2)$$

This is not to be taken literally, since $\mathbf{u}$ represents the ion *relative* displacement field, but $\boldsymbol{\lambda}$ has the usual symmetry properties of the elastic stiffness tensor which stem from general invariance arguments (see section 2). Finally, the first term in Eq. (3.1) is the kinetic energy density while the fifth one is just the energy density in the electric field. The problem is formulated for any general heterostructure with interfaces of arbitrary shape. We also stress that in the formal analysis the constituent media may be isotropic or anisotropic and all quantities entering in Eq. (3.1) have a piecewise dependence on the coordinates. The form of the postulated Lagrangian density results from the above general consideration and describes all the effects one expects to be contained in a model describing a coupling of the **u** and φ fields in a semiconductor nanostructure. We stress that we are not decoupling the longitudinal and transverse parts of the vector **u** and the analysis is in this sense so far quite general.

Standard application of the principle of least action [49] provides the Euler-Lagrange equations of motion

$$\frac{\partial}{\partial t}\frac{\partial \mathcal{L}}{\partial \dot{u}_i} + \sum_{k=1}^{3}\frac{\partial}{\partial x_k}\frac{\partial \mathcal{L}}{\partial u_{i,k}} - \frac{\partial \mathcal{L}}{\partial u_i} = 0 \ , \qquad (3.3)$$

$$\frac{\partial}{\partial t}\frac{\partial \mathcal{L}}{\partial \dot{\varphi}} + \sum_{k=1}^{3}\frac{\partial}{\partial x_k}\frac{\partial \mathcal{L}}{\partial \varphi_k} - \frac{\partial \mathcal{L}}{\partial \varphi} = 0 \ , \quad (i=1,2,3) \qquad (3.4)$$

and the following notations have been used

$$x_1 = x \quad ; x_2 = y \quad ; x_3 = z \ ,$$
$$\dot{u}_i = \tfrac{\partial u_i}{\partial t} \ ; u_{i,k} = \tfrac{\partial u_i}{\partial x_k} \ ; \dot{\varphi} = \tfrac{\partial \varphi}{\partial t} \ ; \varphi_i = \tfrac{\partial \varphi}{\partial x_i} \ . \qquad (3.5)$$

From Eqs. (3.1), (3.3) and (3.4) the following equations of motion are derived

$$\rho \frac{\partial^2 \mathbf{u}}{\partial t^2} - \boldsymbol{\gamma} \cdot \mathbf{u} + \boldsymbol{\alpha} \cdot \nabla\varphi + \nabla \cdot \boldsymbol{\sigma} = 0 \ , \qquad (3.6)$$

$$\nabla \cdot \left(-\boldsymbol{\alpha} \cdot \mathbf{u} + \boldsymbol{\delta} \cdot \nabla\varphi + \frac{1}{4\pi}\nabla\varphi\right) = 0 \ . \qquad (3.7)$$

Equations (3.6) and (3.7) yield four scalar coupled second order partial differential equations for the four quantities **u** and φ with arbitrarily varying coefficients. Note that these coefficients are affected by the differential gradient operator ∇. The fact that $\boldsymbol{\alpha}$ is not affected in (3.6), while it is in (3.7) does not imply any contradiction. The real physical quantity is not the electrostatic potential φ but its gradient, i.e. the electric field. The picture is then clear if (3.6) and (3.7) are read in terms of the real fields **u** and **E**. In (3.6) $\boldsymbol{\alpha}$ is multiplying **E** just as $\boldsymbol{\gamma}$ is multiplying **u**, while (3.7) is just the divergence of the electric displacement

D (Poisson equation), given by (3.11). The formulation in terms of an electrostatic potential in (3.6, 3.7) is a convenient frame for the canonical analysis of section 3.

The momentum field canonically conjugate to **u** is

$$\pi = \frac{\partial \mathcal{L}}{\partial \dot{\mathbf{u}}} = \rho \dot{\mathbf{u}}, \tag{3.8}$$

but there is no momentum conjugate to φ, as this does not appear in the unretarded Lagrangian density. The Hamiltonian density is then

$$\begin{aligned}\mathcal{H} &= \pi \cdot \dot{\mathbf{u}} - \mathcal{L} \\ &= \frac{1}{2\rho}\pi^2 - \frac{1}{2}\mathbf{u}\cdot\gamma\cdot\mathbf{u} - \mathbf{u}\cdot\alpha\cdot\mathbf{E} - \frac{1}{2}\mathbf{E}\cdot\delta\cdot\mathbf{E} \\ &\quad -\frac{1}{8\pi}\mathbf{E}^2 - \frac{1}{2}(\nabla\mathbf{u}):\lambda:(\nabla\mathbf{u}).\end{aligned} \tag{3.9}$$

The Hamiltonian density $\mathcal{H}$ can be related to the free energy density in which **E** is treated as an independent variable, whence the electric induction vector **D** [50]

$$\mathbf{D} = -4\pi \left(\frac{\partial \mathcal{H}}{\partial \mathbf{E}}\right), \tag{3.10}$$

which yields

$$\mathbf{D} = 4\pi\left(\alpha\cdot\mathbf{u} + \frac{1}{4\pi}(\mathbf{E} + 4\pi\delta\cdot\mathbf{E})\right). \tag{3.11}$$

According to the usual definition $\mathbf{D} = \mathbf{E} + 4\pi\mathbf{P}$ this yields a constitutive relation which can also be cast as

$$\mathbf{P} = \alpha\cdot\mathbf{u} + \delta\cdot\mathbf{E}. \tag{3.12}$$

From (3.11) it can be directly seen that (3.7) becomes

$$\nabla\cdot\mathbf{D} = 0, \tag{3.13}$$

expressing the absence of free charges in the medium. Equations (3.6) and (3.7) represent a generalisation of the *Born-Huang equations* [51] for structures formed by inhomogeneous media including dispersion of the oscillations and (3.12) is a generalisation of the *Born-Huang model* valid for arbitrary — e.g. anisotropic — media.

The tensor parameters entering in (3.6) and (3.7) can be expressed in terms of experimentally measurable quantities. By definition of the vector dielectric displacement

$$\mathbf{D} = \epsilon(\omega) \cdot \mathbf{E} \ . \tag{3.14}$$

In the limit $\omega \to \infty$ (for frequencies high compared with the bulk infra-red dispersion frequency or transverse optical frequency ω_T) the mechanical amplitude $\mathbf{u} \to 0$ and (3.11) is reduced to

$$\mathbf{D} = (\mathbf{I} + 4\pi \delta) \cdot \mathbf{E} \ , \tag{3.15}$$

where $\mathbf{I}$ is the unit matrix. Comparing with (3.14) we obtain

$$\delta = \left(\frac{\epsilon_\infty - \mathbf{I}}{4\pi} \right) \ . \tag{3.16}$$

ϵ_∞ being the high-frequency dielectric tensor of the medium. For the static limit $\omega \to 0$ (frequency very low compared with ω_T), $\mathbf{D} = \epsilon_0 \cdot \mathbf{E}$, where ϵ_0 is the static dielectric tensor. Thus we obtain from (3.6) and (3.11)

$$-\gamma \cdot \mathbf{u} - \alpha \cdot \mathbf{E} + \nabla \cdot \sigma = 0 \ , \tag{3.17}$$

$$(\epsilon_0 - \epsilon_\infty) \cdot \mathbf{E} = 4\pi \alpha \cdot \mathbf{u} \ . \tag{3.18}$$

Taking $\nabla \cdot \sigma = 0$ in (3.17) we find that

$$\alpha \cdot \gamma^{-1} \cdot \alpha = \frac{1}{4\pi} (\epsilon_\infty - \epsilon_0) \ . \tag{3.19}$$

Equation (3.19) gives α in terms of directly measurable quantities ϵ_∞, ϵ_0 and γ. The form of the tensor γ can be related to the natural oscillation frequencies of the medium by means of a phenomenological interpretation of its components in equation (3.6). If γ is referred to the principal axes of the bulk medium, then it is diagonal and given by

$$\gamma = -\rho \begin{pmatrix} \omega_{1T}^2 & 0 & 0 \\ 0 & \omega_{2T}^2 & 0 \\ 0 & 0 & \omega_{3T}^2 \end{pmatrix} \ , \tag{3.20}$$

where ω_{iT} ($i = 1, 2, 3$) are the natural frequencies of the medium. The formulae (3.16), (3.19) and (3.20) hold quite generally. In an inhomogeneous system, for instance, the material parameters would depend on position. In heterostructures they are usually taken as piecewise constants, taking the corresponding values in each of the constituent media, which is another way of depending on position. The differential operators are placed in their proper places and act also on position-dependent parameters when they must. Although in this book we shall only deal with piecewise homogeneous heterostructures it is in order to stress that the formulation here presented is also valid if heterogeneities are involved. The solution of the field equations with matching boundary conditions would then obviously be more complicated, but it is still possible to prescribe techniques capable of coping with this situation. For instance, the combination of the Green function analysis presented in Chapter 4 with the *transfer matrix* [52] — a notion which differs from the usual one often employed to study homogeneous systems in that it transfers both, amplitudes and derivatives — yields a technique which can be used to study simultaneously matching and inhomogeneities [53]. Having made this point clear, in the examples discussed in this book we shall always take the heterostructures as piecewise homogeneous.

Starting from (2.9) as in the bulk case, Eq. (3.6) reads

$$\rho\omega^2 \mathbf{u} + \boldsymbol{\gamma} \cdot \mathbf{u} - \boldsymbol{\alpha} \cdot \nabla\varphi - \nabla \cdot \boldsymbol{\sigma} = 0 \,. \tag{3.21}$$

Most of the usually fabricated semiconductor nanostructures are grown with III-V compounds, for which the constituent materials belong to the cubic crystal symmetry. Under these conditions the tensors present in Eqs. (3.16), (3.19) and (3.20) take a simple diagonal form and we obtain:

$$\boldsymbol{\gamma} = -\rho\omega_T^2 \mathbf{I} \,, \tag{3.22}$$

$$\boldsymbol{\delta} = \frac{1}{4\pi}(\epsilon_\infty - 1)\mathbf{I} \,, \tag{3.23}$$

$$\boldsymbol{\alpha} = \sqrt{\frac{(\epsilon_0 - \epsilon_\infty)}{4\pi}\rho\omega_T^2}\,\mathbf{I} = \alpha\mathbf{I} \,. \tag{3.24}$$

The tensor λ then has only there independent components (see Appendix A) which we take as

$$\lambda_1 = \rho\beta_L^2 \;;\; \lambda_2 = \rho(\beta_L^2 - 2\beta_T^2) \;;\; \lambda_3 = \rho\frac{\beta_C^2}{2} \,. \tag{3.25}$$

Thus, the 'stress' tensor is cast in the form

$$\sigma = \rho \begin{pmatrix} \beta_L^2 \frac{\partial u_x}{\partial x} + (\beta_L^2 - 2\beta_T^2)\left(\frac{\partial u_y}{\partial y} + \frac{\partial u_z}{\partial z}\right) & \frac{1}{2}\beta_C^2\left(\frac{\partial u_x}{\partial y} + \frac{\partial u_y}{\partial x}\right) & \frac{1}{2}\beta_C^2\left(\frac{\partial u_x}{\partial z} + \frac{\partial u_z}{\partial x}\right) \\ \frac{1}{2}\beta_C^2\left(\frac{\partial u_x}{\partial y} + \frac{\partial u_y}{\partial x}\right) & \beta_L^2 \frac{\partial u_y}{\partial y} + (\beta_L^2 - 2\beta_T^2)\left(\frac{\partial u_x}{\partial x} + \frac{\partial u_z}{\partial z}\right) & \frac{1}{2}\beta_C^2\left(\frac{\partial u_y}{\partial z} + \frac{\partial u_z}{\partial y}\right) \\ \frac{1}{2}\beta_C^2\left(\frac{\partial u_x}{\partial z} + \frac{\partial u_z}{\partial x}\right) & \frac{1}{2}\beta_C^2\left(\frac{\partial u_y}{\partial z} + \frac{\partial u_z}{\partial y}\right) & \beta_L^2 \frac{\partial u_z}{\partial z} + (\beta_L^2 - 2\beta_T^2)\left(\frac{\partial u_x}{\partial x} + \frac{\partial u_y}{\partial y}\right) \end{pmatrix} \tag{3.26}$$

The phenomenological parameters β_L^2, β_T^2 and β_C^2 describe dispersion in an anisotropic cubic medium and can be estimated from the experimental phonon dispersion relations of the corresponding bulk material. Typical parameters for several semiconductors are listed in Table 1.1. The isotropic case is obtained by just taking $\beta_C^2 = 2\beta_T^2$ and then the σ tensor has only two different components [44] and is reduced to

$$\sigma = \rho\,(\beta_L^2 - \beta_T^2)\,\nabla\cdot\mathbf{u}\,\mathbf{I} + \rho\beta_T^2 \nabla\mathbf{u} \tag{3.27}$$

and Eqs. (3.7) and (3.21) become

$$\rho(\omega^2 - \omega_T^2)\mathbf{u}(\mathbf{r}) = \nabla\cdot(\rho\beta_L^2 \nabla\mathbf{u}) - \nabla\times(\rho\beta_T^2 \nabla\times\mathbf{u}) + \alpha\nabla\varphi \tag{3.28}$$

and

$$\nabla\cdot(\epsilon_\infty \nabla\varphi) = 4\pi\nabla\cdot(\alpha\mathbf{u}) \quad. \tag{3.29}$$

These differential equations, like (3.6) and (3.7), hold for any heterogeneous system or heterostructure with position-dependent coefficients and give the explicit form that (3.6) and (3.7) take for the case of isotropic media.

As in section 2 the physical meaning of (3.29) is that the scalar potential φ is a solution of a generalised Poisson equation where the polarisation charge $\rho_P = \nabla\cdot\mathbf{P}$ of the polarisation field $\mathbf{P}$ is given by Eq. (2.15). A similar physical interpretation can be derived from the general analysis embodied in (3.6) and (3.7). Then the potential φ is created by the polarisation charge $\rho_P = \nabla \cdot [\alpha \cdot \mathbf{u} - (\boldsymbol{\epsilon}_\infty - \mathbf{I})/(4\pi)\nabla\phi]$.

The term $\nabla \cdot (\alpha \mathbf{u})$ in (3.29) measures the effect of the coupling between the φ and $\mathbf{u}$ fields on the equation for the scalar potential while the term $\alpha\nabla\varphi$ measures the effect of the coupling between the $\mathbf{u}$ and φ fields on the equation of motion for $\mathbf{u}$. Hence, the general way to obtain a correct solution is to solve the system of four coupled differential equations [54]. For a bulk medium (isotropic and homogeneous) described by the above equations it is clear that we can obtain independent equation for $\mathbf{u}_L (\nabla \times \mathbf{u}_L = 0)$ and $\mathbf{u}_T (\nabla \cdot \mathbf{u}_T = 0)$ and this yields at once decoupled longitudinal and transverse modes with the dispersion relations (2.16).

However, our concern is to solve Eqs. (3.28) and (3.29) for any arbitrary semiconductor nanostructure without restrictions on its geometry and to study the matching of different media at the interfaces, for which we usually solve (3.28) and (3.29) within a given homogeneous part of the structure and then apply the matching boundary conditions at the said interfaces.

A general method of solution for isotropic homogeneous media

Our concern is to solve (3.28) and (3.29) for heterostructures formed by matching different homogeneous constituent media. A way to do this is to solve the differential system for each constituent (bulk) medium and then to match these solutions by imposing appropriate matching boundary conditions at the interfaces. For the latter we shall find it convenient (section 3) to start from the differential system for the entire inhomogeneous heterostructure, but the first part of the analysis concerns each constituent medium separately.

While real crystals are anisotropic and this feature persists in the continuum long wave limit, it suffices for most cases in practice to make an average isotropic approximation for the constituent media. The effects of crystalline anisotropy will be discussed from a different perspective in section 3. Here we shall discuss a general method with which we can obtain closed analytical solutions for a bulk homogeneous medium.

We introduce two auxiliary functions, namely a scalar Ψ and a vector $\boldsymbol{\Gamma}$ such that

$$\Psi = \nabla \cdot \mathbf{u} \quad ; \quad \boldsymbol{\Gamma} = \nabla \times \mathbf{u} \ . \tag{3.30}$$

Applying the curl operator to Eq. (3.28) in a given homogeneous part of the structure where the medium parameters are constant we obtain

$$(\omega^2 - \omega_T^2)\nabla \times \mathbf{u} = \beta_T^2 \nabla \times \nabla \times \nabla \times \mathbf{u} , \qquad (3.31)$$

where we used the fact that $\nabla \times \nabla f \equiv 0$. Using the vector identity $\nabla \times \nabla \times \mathbf{u} \equiv \nabla \nabla \cdot \mathbf{u} - \nabla^2 \mathbf{u}$ the auxiliary vector $\mathbf{\Gamma}$ satisfies the equation

$$(\nabla^2 + Q_T^2)\mathbf{\Gamma} = 0 , \qquad (3.32)$$

where

$$Q_T^2 = \frac{\omega_T^2 - \omega^2}{\beta_T^2} . \qquad (3.33)$$

Similarly, taking the divergence of Eq. (3.28) we have

$$\rho(\omega^2 - \omega_T^2)\nabla \cdot \mathbf{u} = \rho \beta_L^2 \nabla^2 \nabla \cdot \mathbf{u} + \alpha \nabla^2 \varphi , \qquad (3.34)$$

and according to (3.29) it follows that

$$\alpha \nabla^2 \varphi = \frac{4\pi \alpha^2}{\epsilon_\infty} \nabla \cdot \mathbf{u} . \qquad (3.35)$$

Substituting (3.35) in (3.34) and using (3.24) we obtain the equation for Ψ

$$(\nabla^2 + Q_L^2)\Psi = 0 \qquad (3.36)$$

with

$$Q_L^2 = \frac{\omega_L^2 - \omega^2}{\beta_L^2} . \qquad (3.37)$$

The above equations are decoupled with respect to Ψ and $\mathbf{\Gamma}$ and hence it is straightforward to solve them. On comparing (3.35) and (3.36) it can be seen that the general solution for the scalar potential is given by

$$\boxed{\varphi = \varphi_H - \frac{4\pi \alpha}{\epsilon_\infty Q_L^2} \Psi ,} \qquad (3.38)$$

where φ_H satisfies the Laplace equation

$$\nabla^2 \varphi_H = 0 \qquad (3.39)$$

and the term $4\pi\alpha/(\epsilon_\infty Q_L^2)\,\Psi$ in (3.38) is a particular solution of (3.36). The general solution for the vibrational displacement $\mathbf{u}$ can then be obtained by replacing Eqs. (3.30) and (3.38) into Eq. (3.28). Thus

$$\mathbf{u} = -\nabla\left(\frac{\alpha}{\rho\beta_T^2 Q_T^2}\varphi_H + \frac{\Psi}{Q_L^2}\right) + \frac{1}{Q_T^2}\nabla\times\mathbf{\Gamma}. \qquad (3.40)$$

Once we have obtained the general solution of the Helmholtz equations (3.32) and (3.36), and of the Laplace equation (3.39) for φ_H we obtain $\mathbf{u}$ and φ by means of (3.38) and (3.40), respectively. With this procedure we obtain general analytical solutions for the coupled quantities $\mathbf{u}$ and φ which can be used to study any given heterostructure by combining with the corresponding matching conditions. This determines the final solution.

Finally, it can be seen from (3.40) that the vibrational displacement can be written as a sum of two non independent terms: namely a longitudinal part $\mathbf{u}_L$ given by

$$\mathbf{u}_L = -\nabla\left[\frac{\alpha}{\rho\beta_T^2 Q_T^2}\varphi_H + \frac{\Psi}{Q_L^2}\right], \qquad (3.41)$$

and a transverse one $\mathbf{u}_T$ which has the form

$$\mathbf{u}_T = \frac{1}{Q_T^2}\nabla\times\mathbf{\Gamma}. \qquad (3.42)$$

We stress that in general $\mathbf{u}_L$ and $\mathbf{u}_T$ cannot be decoupled for heterostructures involving interfaces, a point to which we shall return in Chapter 4.

Matching conditions

We now turn to the question of boundary conditions and distinguish between two concepts. One is that in *any* differential system we must specify the *asymptotic boundary conditions* and the other one is that a matching problem entails certain matching rules, which we shall term *matching boundary conditions* to distinguish them from the former.

The asymptotic *boundary conditions* are

$$|\mathbf{u}| < \infty, \quad |\varphi| < \infty, \quad \forall\,\mathbf{r}, \qquad (3.43)$$

since $\mathbf{u}$ and φ must be bounded everywhere as they describe physical quantities.

The *matching boundary conditions* follow directly from the equations of motion for the entire system. Let us assume that we have two different media

Polar optical modes in heterostructures 33

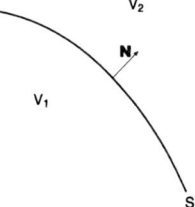

Figure 3.1 Two different media, in domains of Volume V_1 and V_2, meet at the surface S where matching is to be effected. The figure shows the sign convention for the normal unit vector **N**.

labelled 1 and 2 enclosed in volumes V_1 and V_2 separated by a surface S with normal unit vector **N** pointing into medium 2. (See Figure 3.1).

We operate with partial differential equations for **u** and φ, and therefore any functions **u** and φ must be continuous in the whole space and in particular at the surface S, whence a first set of matching conditions:

$$\mathbf{u}|_{r \in S_-} = \mathbf{u}|_{r \in S_+} \quad , \varphi|_{r \in S_-} = \varphi|_{r \in S_+} \; . \tag{3.44}$$

The symbols S_+ (S_-) indicate that the corresponding functions are evaluated on side 2 (1) of S. Given that the parameters entering the basic equations (3.6) and (3.7) have a piecewise dependence or **r** and change abruptly at the interface, the partial derivatives of **u** and φ need not be continuous at S. Equation (3.6) contains $\nabla \cdot \sigma$ while Poisson's equation (3.7) contains $\nabla \cdot \mathbf{D}$ and the discontinuities of the normal derivatives of **u** and φ at S can be determined from the equations. Integration of Eqs. (3.6) and (3.7) over a volume V surrounding S and shrinking upon it yields for all $\mathbf{r} \in S$, after application of the Gauss theorem, the second set of matching boundary conditions

$$\sigma \cdot \mathbf{N}|_{r \in S_-} = \sigma \cdot \mathbf{N}|_{r \in S_+} \; , \tag{3.45}$$

$$\mathbf{D} \cdot \mathbf{N}|_{r \in S_-} = \mathbf{D} \cdot \mathbf{N}|_{r \in S_+} \; . \tag{3.46}$$

These matching rules have a clear physical interpretation. Equation (3.45) expresses the continuity of the force flux across the interface while, (3.46) is the continuity of the normal component of the electric displacement, D_N. Equations (3.43)–(3.46) are mathematical conditions which must fulfil the functions **u** and φ in order to be solutions of the field equations in the entire space. We emphasise that the above matching boundary conditions are not artificially imposed after invoking intuitive physical laws, but derived from the field equations in a formal way.

We next consider particular situations involving more restrictive matching conditions. Suppose the condition of continuity of **u** Eq. (3.44) at the interface is substituted by

$$\mathbf{u}|_{\mathbf{r}\in S} = 0 \quad . \tag{3.47}$$

This assumes that the interface is a rigid wall for the mechanical vibration. Here, the continuity of φ at $\mathbf{r} \in S$ must still be imposed but the condition (3.46) can be reduced to

$$(\boldsymbol{\epsilon}_\infty \cdot \nabla\varphi) \cdot \mathbf{N}|_{\mathbf{r}\in S_-} = (\boldsymbol{\epsilon}_\infty \cdot \nabla\varphi) \cdot \mathbf{N}|_{\mathbf{r}\in S_+} \quad . \tag{3.48}$$

The latter condition is derived from the continuity of D_N and Eq. (3.11) according to condition (3.23) and Eq. (3.47). These simplified matching conditions can be justified from the large separation between the optical branches of the two components. For example, in GaAs/AlAs structures the AlAs side of the structure does not vibrate mechanically in the range of the GaAs oscillations and the bandwidths of the GaAs and AlAs polar optical phonons are much smaller than the separation between them. Thus, AlAs behaves as a completely rigid medium for the **u** vibrations of GaAs which do not penetrate significantly into AlAs. In this case Newton's law is defined only in the domain $\mathbf{r} \in (V_1 + S)$, while Poisson's equation (3.7) is defined everywhere $\mathbf{r} \in (V_1 + V_2 + S)$. That is, if one studies the GaAs modes in GaAs/AlAs structures, then Eq. (3.6) is defined in the domain containing GaAs while the electrostatic potential is defined in the entire structure. It is clear from the condition (3.47) that the 'stress' boundary condition (3.45) becomes unnecessary in this case.

Another case is that of *free oscillations* where the active medium is surrounded by vacuum, thus admitting free oscillations at the surface. In this case the matching boundary conditions are:

- continuity of φ at the surface,
- $\boldsymbol{\sigma} \cdot \mathbf{N}|_{\mathbf{r}\in S} = 0$,
- continuity of D_N at the surface

The latter condition entails the following relation

$$(4\pi\boldsymbol{\alpha} \cdot \mathbf{u} + \boldsymbol{\epsilon} \cdot \mathbf{E})|_{\mathbf{r}\in S_-} = \boldsymbol{\epsilon}\mathbf{E}|_{\mathbf{r}\in S_+} \quad . \tag{3.49}$$

The vector **u** is then defined in just one side of the structure, that is in the active medium. For a free surface Poisson's equation is defined everywhere but (3.6) is only defined for $\mathbf{r} \in (V_1 + S)$.

The extension of the formal argument to more than one interface is immediate.

Having seen the general arguments concerning both, the basic equations of the phenomenological model and the matching boundary conditions it is interesting to pause and consider a technical point which is at the root of many difficulties encountered in the literature.

The point is the following: Consider a general vector field **V** with longitudinal ($\mathbf{V}_L$; $\nabla \times \mathbf{V}_L = 0$) and transverse ($\mathbf{V}_T$; $\nabla \cdot \mathbf{V}_T = 0$) parts. Then the matching boundary conditions mix up longitudinal and transverse components [55, 56] so that there is no purely longitudinal solution satisfying the matching boundary conditions, *even if in the bulk media we start from purely longitudinal fields* [56]. Yet the assumption that heterostructures have longitudinal optical modes — for which **u** is a purely longitudinal vector $\mathbf{u}_L$ — is often made. Then, if this assumption is made, $\nabla \times \mathbf{u}$ in (3.28) vanishes and, since **E** in the quasistatic limit is always longitudinal and so is **u** now by assumption, we find a proportionality relationship between the two which, in Fourier transform, reads

$$\rho(\omega^2 - \omega_T^2 - \beta_L^2 k^2)\mathbf{u} + \alpha \mathbf{E} = \mathbf{0}. \tag{3.50}$$

Then **P**, given in the isotropic case by

$$\mathbf{P} = \alpha \mathbf{u} + \frac{\epsilon_\infty - 1}{4\pi}\mathbf{E} \tag{3.51}$$

is also longitudinal and so is **D**, which is $\mathbf{E} + 4\pi \mathbf{P}$. Thus **D** has no transverse part. But, with no external charge to create a longitudinal **D** field, **D** has no longitudinal part either. Thus with these assumptions **D** *vanishes identically*.

Having reached this stage one has two options. One is to put $\mathbf{P} = -\mathbf{E}/4\pi$. Then, after eliminating **P** between this and (3.51) one has

$$\mathbf{P} = -\frac{4\pi \alpha}{\epsilon_\infty}\mathbf{u} \tag{3.52}$$

and then, on using this to eliminate **E** from (3.50) and resorting to the *Lydane-Sachs-Teller relation* we obtain an equation of motion for **u** *alone*, namely

$$\rho\left(\omega^2 - \omega_L^2 + \beta_L^2 k^2\right) = 0. \tag{3.53}$$

So far this is correct in bulk media and yields the well-known dispersion relation for the longitudinal optical modes. What is incorrect is to assume that all the assumptions made hold also for a matched system. This is the basis of the models of a purely mechanical type in which (3.53) is written in 2D Fourier transform — with 2D wavevector κ — as an ordinary differential equation for u_y, i.e.

$$\left[\frac{d^2}{dz^2} + \frac{\omega_L^2 - \omega^2}{\beta_L^2} - \kappa^2\right] u_y = 0 . \tag{3.54}$$

(The vector κ has been taken in the y direction and the condition $\nabla \times \mathbf{u} = 0$ has been used to eliminate u_x).

The differential equation (3.54) is then taken as the starting point to effect the matching at the interfaces by imposing *only mechanical* matching conditions and it is not surprising to find at the end an electrostatic discontinuity at the matching surfaces.

The other alternative is to write $\mathbf{u}$ from (3.50) in terms of $\mathbf{E}$ and use the constitutive relation (3.51) to eliminate $\mathbf{u}$. Then one finds an expression for $\mathbf{P}$ solely in terms of $\mathbf{E}$, whence a relationship of the form $\mathbf{D} = \epsilon(\omega, k)\mathbf{E}$ with a *longitudinal* dielectric function

$$\epsilon(\omega, k) = \epsilon_\infty \frac{\omega^2 - \omega_L^2 + \beta_L^2 k^2}{\omega^2 - \omega_T^2 + \beta_T^2 k^2} . \tag{3.55}$$

This is the basis of the *dielectric models* in which (3.55) describes the constituent bulk media and standard dielectric matching is effected. This cures the dielectric discontinuity but, not surprisingly, introduces a *mechanical* discontinuity.

The only way to perform a formally correct matching analysis is to accept that $\mathbf{u}$ has longitudinal and transverse parts, to start from the coupled differential equations (3.6) and (3.7) and to impose all the matching boundary conditions. More will be said about this in Chapter 4.

Hermiticity, orthogonality and completeness

We have four amplitudes $(\mathbf{u}, \varphi)$ and four coupled second order differential equations for $\mathbf{u}$ and φ. It is convenient to define a tetrafield column-matrix

$$\mathbf{F} = \begin{bmatrix} \mathbf{u} \\ \varphi \end{bmatrix} \tag{3.56}$$

which has mechanical and electrical components and the basic equations formed by the coupled Euler-Lagrange Eqs. (3.6) and (3.7), can then be cast in matrix form

$$\hat{L} \cdot \mathbf{F} = \rho\omega^2 \begin{pmatrix} I & 0 \\ 0 & 0 \end{pmatrix} \mathbf{F} . \tag{3.57}$$

Polar optical modes in heterostructures 37

The linear operator $\hat{\mathbf{L}}$ is a 4×4 differential matrix

$$\hat{\mathbf{L}} = \begin{pmatrix} \hat{L}_{MM} & \hat{L}_{ME} \\ \hat{L}_{EM} & \hat{L}_{EE} \end{pmatrix} , \tag{3.58}$$

where

$$\begin{aligned} \hat{L}_{MM} \cdot \mathbf{u} &= -\boldsymbol{\gamma} \cdot \mathbf{u} + \nabla \cdot \boldsymbol{\sigma} , \\ \hat{L}_{ME}\, \varphi &= \boldsymbol{\alpha} \cdot \nabla \varphi , \\ \hat{L}_{EM} \cdot \mathbf{u} &= -\nabla \cdot [\boldsymbol{\alpha} \cdot \mathbf{u}] , \\ \hat{L}_{EE}\, \varphi &= \tfrac{1}{4\pi} \nabla \cdot [(\mathbf{I} + 4\pi\,\boldsymbol{\delta}) \cdot \nabla \varphi] , \end{aligned} \tag{3.59}$$

with M referring to mechanical (vibrational) and E to the electrical parts. This condenses the system (3.6), (3.7) in compact form as

$$\begin{aligned} \hat{L}_{MM} \cdot \mathbf{u} + \hat{L}_{ME}\, \varphi &= \rho\omega^2 \mathbf{u} \\ \hat{L}_{EM} \cdot \mathbf{u} + \hat{L}_{EE}\, \varphi &= 0 . \end{aligned} \tag{3.60}$$

Let us now determine the necessary and sufficient conditions for the hermiticity of operator $\hat{\mathbf{L}}$. Consider any pair of functions $\mathbf{F}_i$ and $\mathbf{F}_j$ of the tetrafield space and study the general matrix elements

$$L_{ij} = \int_V \mathbf{F}_i^\dagger \cdot \hat{\mathbf{L}} \cdot \mathbf{F}_j\, dV , \tag{3.61}$$

where V is the integration volume for the entire system. By using the definition (3.58), (3.59) of $\hat{\mathbf{L}}$ acting on $\mathbf{F}_j$ and the constitutive relation (3.11) we are led to

$$L_{ij} = \int_V \left[-\mathbf{u}_i{}^* \cdot \boldsymbol{\gamma} \cdot \mathbf{u}_j + \mathbf{u}_i{}^* \cdot \nabla \cdot \boldsymbol{\sigma}_j + \mathbf{u}_i{}^* \cdot \boldsymbol{\alpha} \cdot \nabla \varphi_j - \frac{1}{4\pi} \varphi_i^* \nabla \cdot \mathbf{D}_j \right] dV . \tag{3.62}$$

Here $\boldsymbol{\sigma}_j$ and $\mathbf{D}_j$ are the 'stress' tensor and electric displacement vector associated with $\mathbf{u}_j, \varphi_j$ according to Eqs. (3.2) and (3.11), respectively. Consider now the case of a system consisting of two different media with volumes V_1 and V_2 matched at a surface S. The functions $\nabla \cdot \boldsymbol{\sigma}_i$ and $\nabla \cdot \mathbf{D}_j$ contain singular terms going like $\pm\delta(\mathbf{r}|_{\mathbf{r}\in S} - \mathbf{r})\mathbf{N}$. By noting that

$$\nabla \cdot (\mathbf{u}_i \cdot \boldsymbol{\sigma}_j) = \nabla \mathbf{u}_i : \boldsymbol{\sigma}_j + \mathbf{u}_i \cdot \nabla \boldsymbol{\sigma}_j \qquad (3.63)$$

and

$$\nabla \cdot (\varphi_i^* \mathbf{D}_j) = \nabla \varphi_i^* \cdot \mathbf{D}_j + \varphi_i^* \nabla \cdot \mathbf{D}_j \qquad (3.64)$$

the matrix elements L_{ij} can be rewritten as

$$L_{ij} = A_{ij} + B_{ij}, \qquad (3.65)$$

with

$$A_{ij} = \int_V \{-\mathbf{u}_i^* \cdot \boldsymbol{\gamma} \cdot \mathbf{u}_j - (\nabla \mathbf{u}_i^*) : \boldsymbol{\lambda} : \nabla \mathbf{u}_j + \mathbf{u}_i^* \cdot \boldsymbol{\alpha} \cdot \nabla \varphi_j \\ + \mathbf{u}_j \cdot \boldsymbol{\alpha} \cdot \nabla \varphi_i^* + 1/(4\pi)(\nabla \varphi_i^*) \cdot \nabla \varphi_j + (\nabla \varphi_i^*) \cdot \boldsymbol{\delta} \cdot \nabla \varphi_j\} \, dV, \qquad (3.66)$$

$$B_{ij} = \int_V \left[-\nabla \cdot (\varphi_i^* \mathbf{D}_j) - \nabla \cdot (\mathbf{u}_i^* \cdot \boldsymbol{\sigma}_j) \right] dV, \qquad (3.67)$$

where Eqs. (3.2) and (3.11) have been used. It follows from the general symmetries of the tensors $\boldsymbol{\gamma}$, $\boldsymbol{\lambda}$ and $\boldsymbol{\delta}$ involved that $A_{ij} = A_{ji}^*$. Thus the difference between the matrix elements L_{ij} and L_{ji}^* lies only in $B_{ij} - B_{ji}^*$ where all the singular parts are contained. Now, the volume of integration in (3.67) consists of V_1 and V_2 and this yields

$$B_{ij} = \\ \int_S \mathbf{u}_i^* \cdot \left(\boldsymbol{\sigma}_j|_{\mathbf{r}\in S_+} - \boldsymbol{\sigma}_j|_{\mathbf{r}\in S_-} \right) \cdot \mathbf{N} \, dS \\ + \int_S \varphi_i^* \left(\mathbf{D}_j|_{\mathbf{r}\in S_+} - \mathbf{D}_j|_{\mathbf{r}\in S_-} \right) \cdot \mathbf{N} \, dS. \qquad (3.68)$$

By using (3.65)-(3.68) we are led to the following expression:

$$L_{ij} = L_{ji}^* + \int_S \left\{ \mathbf{u}_i^* \cdot \left(\boldsymbol{\sigma}_j^{(+)} - \boldsymbol{\sigma}_j^{(-)} \right) \cdot \mathbf{N} - \mathbf{u}_j^* \left(\boldsymbol{\sigma}_i^{(+)} - \boldsymbol{\sigma}_i^{(-)} \right) \cdot \mathbf{N} \right. \\ \left. + \varphi_i^* \left(\mathbf{D}_j^{(+)} - \mathbf{D}_j^{(-)} \right) \cdot \mathbf{N} - \varphi_j \left(\mathbf{D}_i^{(+)} - \mathbf{D}_i^{(-)} \right)^* \cdot \mathbf{N} \right\} dS. \qquad (3.69)$$

Thus the necessary and sufficient condition, ensuring the hermiticity of the operator $\hat{L}$ is that for all $\mathbf{F}_i$, $\mathbf{F}_j$ the surface terms vanish. This allows us to identify the space of solutions which satisfy this condition. There are several obvious situations ensuring the hermiticity of the matrix L_{ij}, being compatible with the equations of this theory, namely:

(i) The trivial case of a bulk continuous with $\mathbf{u}$, φ and its derivatives continuous

(ii) The functions satisfying the matching boundary conditions (3.44)-(3.46)

(iii) Matching boundary conditions with complete confinement: $\mathbf{u} = 0$ for $\mathbf{r} \in S$, continuity of φ and D_N at S

(iv) Matching boundary conditions for a free surface: $\boldsymbol{\sigma} \cdot \mathbf{N} = 0$ for $\mathbf{r} \in S$, continuity φ and D_N at S. In this case the Poisson equation is defined everywhere but the equation for $\mathbf{u}$ is only defined in the domain filled by the material medium.

The extension of the formal argument to more than one interface is immediate. These conclusions hold for any structure having more than one interface and arbitrary geometry. As can be seen from (3.57) we have an eigenvalue problem. Let m be a generic label for the different eigenstates $\mathbf{F}_m$ obtained after solution of Eq. (3.57) with the eigenfrequencies ω_m. The m-th 'eigenvector' $\mathbf{F}_m$ consists of $\mathbf{u}_m$ and φ_m. The term 'vector' is here used in the sense of a functional space. Only the three components of the $\mathbf{u}$ part of $\mathbf{F}$ transform among themselves as the components of a vector in the standard geometrical sense under orthogonal transformations of coordinates. Now, owing to the nature of Eq. (3.57), the φ component of the eigenvector $\mathbf{F}_m$ always displays a zero eigenvalue. Therefore the properties of orthogonality and completeness of the solutions involve only just the vector $\mathbf{u}_m$. It is easy to prove that

$$\mathbf{F}_m^\dagger \cdot \hat{\mathbf{L}} \cdot \mathbf{F}_n - \mathbf{F}_n^\dagger \cdot \hat{\mathbf{L}}^\dagger \cdot \mathbf{F}_m = \rho \left[\omega_n^2 - \omega_m^{2*} \right] \mathbf{u}_m^* \cdot \mathbf{u}_n . \quad (3.70)$$

Performing the integral over the whole space of Eq. (3.70) we obtain:

$$L_{mn} - L_{nm}^* = \left(\omega_n^2 - \omega_m^{2*} \right) \int_V \rho \, \mathbf{u}_m^* \cdot \mathbf{u}_n \, dV . \quad (3.71)$$

Hence, due to the hermiticity of $\mathbf{L}$ the eigenvalues ω_m are real quantities and for $m \neq n$ we have

$$\int_V \rho(\mathbf{r}) \mathbf{u}_m^*(\mathbf{r}) \cdot \mathbf{u}_n(\mathbf{r}) dV = 0 . \quad (3.72)$$

This condition represents the orthogonality property of the vector set $\{\mathbf{u}_m\}$, where the functions φ_m are not involved. Note that ρ enters in (3.72) as a weight factor.

The hermiticity of the operator $\hat{L}$ ensures that the $\{u_m\}$ constitute a complete set of eigenfunctions of the problem described by (3.57). Hence, *if we so choose*, the solutions of the present problem can be described by means of a complete ortho*normal* basis of eigenvectors $\{u_m\}$ satisfying the orthonormality conditions

$$\int_V \rho(\mathbf{r})\mathbf{u}_m^*(\mathbf{r}) \cdot \mathbf{u}_n(\mathbf{r}) \, dV = \delta_{nm} . \qquad (3.73)$$

Note that since the right hand side of (3.73) is dimensionless, the vector $\mathbf{u}_m$ according to this should have dimensions $[M^{-1/2}]$. Since the differential equations (3.60) are homogeneous, all components of $\mathbf{F}_m$ are undefined up to an arbitrary factor which fixes the normalisation of the eigenvectors $\{u_m\}$. Thus, we can choose this so that the orthonormality of the $\mathbf{u}_m$ reads

$$\int_V \rho(\mathbf{r})\mathbf{u}_m^*(\mathbf{r}) \cdot \mathbf{u}_n(\mathbf{r}) \, dV = \Omega_0 \delta_{nm} . \qquad (3.74)$$

The dimensions of the arbitrary constant Ω_0 determine these of the $\mathbf{u}_m$ and hence also the φ_m. We can choose this so that Ω_0 has dimensions $[M \, L^2]$. Then the $\mathbf{u}_m$ have the dimensions of a length and the φ_m have those of an electrostatic potential.

The completeness of the vector basis $\{u_m\}$ is expressed by the following relation:

$$\sum_m \rho(\mathbf{r})\mathbf{u}_m^*(\mathbf{r}) \cdot \mathbf{u}_m(\mathbf{r}') = \delta(\mathbf{r} - \mathbf{r}') \qquad (3.75)$$

if (3.73) is chosen, or

$$\sum_m \rho(\mathbf{r})\mathbf{u}_m^*(\mathbf{r}) \cdot \mathbf{u}_m(\mathbf{r}') = \Omega_0 \delta(\mathbf{r} - \mathbf{r}') . \qquad (3.76)$$

if (3.74) is chosen.

We stress that these considerations hold quite generally for any heterostructure and their observance ensures the orthogonality and completeness of the solution. It should be noted that the correct solution space for the present problem is obtained whenever the coupled character of the equations of the theory is properly observed. A failure to do so can be seen to lie at the root of many difficulties encountered with several phenomenological models which have been tried for this problem [54].

A perturbative treatment of crystal anisotropy

Most of the analysis so far expounded holds quite generally for arbitrary constituent media. Nothing in the derivation of the matching boundary conditions, for instance, or in the canonical analysis of section 3, requires these media to be isotropic. The effects of crystalline anisotropy do show up experimentally, both for bulk media and for matched heterostructures, even in the long wave limit. This can be seen, for instance, in microscopic models [22] when one considers angular dispersion in different in-plane directions. For (110) GaAs/AlAs superlattices the microscopic calculations show that for angular dispersion between the [110] and [1$\bar{1}$0] directions the odd confined modes anticross with the antisymmetric interface modes, whereas the even confined modes are unaffected [8, 57].

There is no basic difficulty in accounting for crystal anisotropy — in the long wave continuum model — in an exact manner, but then in many cases, depending on the problem under study, most of the work amounts to numerical computation. In practice, there are many situations in which the effect of crystal anisotropy is actually small, so it makes sense to see how this can be treated as a perturbation. We now discuss a method which allows us to make some progress in analytical terms. This can reduce substantially the numerical computation and in this respect a comment is in order. The unperturbed solutions from which we start the perturbation analysis could be, if we wanted, those of the separate constituent media. While this would seem the natural way to proceed, it would not be the most practical one, for the subsequent matching analysis would start from the perturbed — i.e. more complicated — constituent media. It is more practical to first carry out the matching analysis for simpler, isotropic constituents and then to introduce in the solutions for the matched system the perturbation due to crystal anisotropy.

The anisotropic correction arises from using the complete form of the tensor σ given by (3.27). We recall that for isotropic media $\beta_C^2 = 2\beta_T^2$, so we write the tensor σ as

$$\sigma = \sigma_0 + \sigma_1 \qquad (3.77)$$

where σ_0 is given by Eq. (3.27) and σ_1 is

$$\sigma_1 = -\frac{\rho}{2}(\beta_C^2 - 2\beta_T^2) \begin{pmatrix} 0 & (\frac{\partial u_x}{\partial y} + \frac{\partial u_y}{\partial x}) & (\frac{\partial u_x}{\partial z} + \frac{\partial u_z}{\partial x}) \\ (\frac{\partial u_x}{\partial y} + \frac{\partial u_y}{\partial x}) & 0 & (\frac{\partial u_y}{\partial z} + \frac{\partial u_z}{\partial y}) \\ (\frac{\partial u_x}{\partial z} + \frac{\partial u_z}{\partial x}) & (\frac{\partial u_y}{\partial z} + \frac{\partial u_z}{\partial y}) & 0 \end{pmatrix}.$$
(3.78)

The equation of motion (3.21) can thus be separated into a part $\hat{L}_0$, which is isotropic, plus an anisotropic correction $\hat{L}_1$. Hence we have an operator $\hat{L} = \hat{L}_0 + \hat{L}_1$, where

$$\hat{L}_0 \mathbf{F} = \omega_T^2 \mathbf{u} + \nabla \cdot \boldsymbol{\sigma}_0 + \alpha \nabla \varphi \tag{3.79}$$

and

$$\hat{L}_1 \mathbf{F} = \nabla \cdot \boldsymbol{\sigma}_1 . \tag{3.80}$$

Note that the operator $\hat{L}_1$ is proportional to $(\beta_C^2 - 2\beta_T^2)/2$.

In several semiconductors the σ_1 term is small compared with σ_0 and can be treated as a perturbation to the isotropic solutions (3.38) and (3.40). For example, the dispersion along (110) in GaAs for the two transverse optical vibrations are described by $\frac{1}{2}\beta_C^2$ and β_T^2 and $(\beta_C^2 - 2\beta_T^2)/2$ is a small correction to β_T^2.

With $\hat{L}_0$ as the unperturbed operator and $\hat{L}_1$ as the perturbation, the solution of (3.57) can be written as

$$\mathbf{F} = \sum_m C_m \mathbf{F}_m^{(0)} . \tag{3.81}$$

$\{\mathbf{F}_m^{(0)}\}$ is the complete set of solutions of the isotropic equation, that is a complete set of functions for the operator $\hat{L}_0$, and the C_m are constants.

According to Eqs. (3.81) and (3.57) the equation of motion is cast as

$$\sum_m C_m [\hat{L}_0 \mathbf{F}_m^{(0)} + \hat{L}_1 \mathbf{F}_m^{(0)}] = \sum_m C_m \rho \omega^2 \begin{pmatrix} I & 0 \\ 0 & 0 \end{pmatrix} \mathbf{F}_m^{(0)} . \tag{3.82}$$

Multiplying (3.82) by $\mathbf{F}_n^{(0)*}$ and using the orthonormality condition (3.73) we obtain

$$\sum_m C_m [\rho \omega_m^{(0)2} \delta_{mn} < |\mathbf{u}_m^{(0)}|^2 > + < \mathbf{F}_n^{(0)} |\hat{L}_1| \mathbf{F}_m^{(0)} >] =$$

$$\sum_m \delta_{mn} < |\mathbf{u}_m^{(0)}|^2 > \rho\omega^2 C_m , \qquad (3.83)$$

where $\omega_m^{(0)}$ are the frequencies of the corresponding isotropic case,

$$< |\mathbf{u}_m^{(0)}|^2 > = \int_V |\mathbf{u}_m^{(0)}|^2 \, dV \qquad (3.84)$$

and

$$< \mathbf{F}_n^{(0)} | \hat{\mathbf{L}}_1 | \mathbf{F}_m^{(0)} > = \int_V \mathbf{u}_n^{(0)*} \nabla \cdot \boldsymbol{\sigma}_1 \mathbf{u}_m \, dV . \qquad (3.85)$$

According to (3.84) and (3.85) Eq. (3.83) becomes

$$\sum_m C_m [\rho \omega_m^{(0)2} \delta_{mn} < |\mathbf{u}_m^{(0)}|^2 > + < \mathbf{u}_n^{(0)} | \nabla \cdot \boldsymbol{\sigma}_1 | \mathbf{u}_m^{(0)} >] = \sum_m \rho\omega^2 C_m \delta_{mn} < |\mathbf{u}_m^{(0)}|^2 > . \qquad (3.86)$$

Taking the parameter $\lambda = \rho(\beta_C^2 - 2\beta_T^2)/2$ as a small parameter and expanding the coefficients C_m and ω^2 in terms of the perturbation λ

$$C_m = C_m^{(0)} + \lambda C_m^{(1)} + \lambda^2 C_m^{(2)} + \dots , \qquad (3.87)$$

$$\Delta\omega^2 \equiv \omega^2 - \omega^{(0)2} = f_1 \lambda + f_2 \lambda^2 + \dots , \qquad (3.88)$$

the frequency correction due to the anisotropic part of the tensor $\boldsymbol{\sigma}_1$ can be obtained up to different orders in λ.

By equating first order terms in λ the frequency correction is given by

$$\Delta\omega_m^2 = \frac{< \mathbf{u}_m^{(0)} | \nabla \cdot \boldsymbol{\sigma}_1 | \mathbf{u}_m^{(0)} >}{< |\mathbf{u}_m^{(0)}|^2 >} . \qquad (3.89)$$

Thus the frequency of the m-th mode is given by

$$\omega_m^2 = \omega_m^{(0)2} + \frac{< \mathbf{u}_m^{(0)} | \nabla \cdot \boldsymbol{\sigma}_1 | \mathbf{u}_m^{(0)} >}{< |\mathbf{u}_m^{(0)}|^2 >} . \qquad (3.90)$$

We stress that we need to determine the value of the dispersion parameters β_L, β_T and β_C. Normally we use the experimentally measured dispersion of bulk semiconductor along different directions in the Brillouin Zone and expressions for the longitudinal optical and transversal optical phonon frequencies along the selected directions. In conclusion we have in the framework of a continuum theory a perturbative approach to calculate analytic expressions for the effect of anisotropy on optical vibrations in nanostructures. The model can be used to treat modes with wavevectors along a general direction in a quantum well, quantum well wire or quantum dot as well as for nanostructures grown along a general direction.

Electron-phonon interaction Hamiltonian

Let us now discuss a general formulation of the electron-phonon Hamiltonian for a Fröhlich-like interaction in heterostructures. Since $\hat{L}$ is Hermitean its normal modes form a complete orthonormal basis, thus we can construct a general displacement field $\mathbf{u}(\mathbf{r}, t)$ by a linear superposition of eigenvectors $\mathbf{u}_n$. In a similar way we can construct a general potential field $\varphi(\mathbf{r}, t)$.

That is

$$\mathbf{u}(\mathbf{r}, t) = \sum_m \left[\bar{C}_m \mathbf{u}_m(\mathbf{r}) e^{-i\omega_m t} + C.C. \right] \tag{3.91}$$

and

$$\varphi(\mathbf{r}, t) = \sum_m \left[\bar{C}_m \varphi_m(\mathbf{r}) e^{-i\omega_m t} + C.C. \right] . \tag{3.92}$$

Compare (3.91) with (2.25). In the present case m stands for *all* the labels of the normal modes of the heterostructure under study. For instance, in the case of planar heterostructures — e.g. a quantum well — these would include a branch number n and a 2D wavevector κ (section 5). Then the vectors $\mathbf{u}_m(\mathbf{r})$ are of the form (5.4), which displays a 2D plane wave in the (x, y) plane and z-dependent function.

The constants $\bar{C}_m$ are the same because they describe the solutions of the eigenvalue problem (3.57). Note that once $\mathbf{u}_m$ is known, φ_m is uniquely determined as $\mathbf{F}_m$ is a solution of a system of four homogeneous linear differential equations. Thus $\varphi(\mathbf{r}, t)$ is known once $\mathbf{u}(\mathbf{r}, t)$ is known.

The momentum density, according to Eq. (3.8), is given by

$$\pi(\mathbf{r}, t) = -i\rho \sum_m \omega_m \left[\bar{C}_m \mathbf{u}_m(\mathbf{r}) e^{-i\omega_m t} - C.C. \right] . \tag{3.93}$$

The canonical quantum expressions for $\mathbf{u}$ and φ are obtained by means of the formal substitution

$$\bar{C}_m \to C_m \hat{b}_m \quad ; \quad \bar{C}_m^* \to C_m \hat{b}_m^\dagger , \tag{3.94}$$

Polar optical modes in heterostructures

where C_m is taken to be real, and $\hat{b}_m$ and $\hat{b}_m^\dagger$ are second-quantisation Bose operators obeying the commutation relations:

$$[\hat{b}_m, \hat{b}_n] = [\hat{b}_m^\dagger, \hat{b}_n^\dagger] = 0, \qquad [\hat{b}_m, \hat{b}_n^\dagger] = \delta_{mn} . \tag{3.95}$$

Obviously, $\hat{b}_m$ and $\hat{b}_m^\dagger$ are the annihilation and creation operators, respectively for the optical oscillation in the state m.

The Fröhlich type electron-polar optical oscillation interaction Hamiltonian is

$$\hat{H}_F = e\hat{\varphi}(\mathbf{r}, t) \tag{3.96}$$

and for this the amplitudes C_m present in φ and $\mathbf{u}$ must be determined. The normalisation of the eigenvectors $\mathbf{u}_m$ is chosen by convention according to Eqs. (3.73) or (3.74), *but the determination of the amplitude coefficients C_m is not optional*. This can be obtained by application of the commutation rule between the $\hat{\mathbf{u}}$ and $\hat{\pi}$ operators

$$[\hat{\mathbf{u}}(\mathbf{r}, 0), \hat{\pi}(\mathbf{r}', 0)] = i\hbar\delta(\mathbf{r} - \mathbf{r}') , \tag{3.97}$$

where

$$\hat{\mathbf{u}}(\mathbf{r}, t) = \sum_m C_m \left[\mathbf{u}_m(\mathbf{r}) e^{-i\omega_m t} \hat{b}_m + H.C. \right] \tag{3.98}$$

and

$$\hat{\pi}(\mathbf{r}, t) = -i\rho \sum_m \omega_m C_m \left[\mathbf{u}_m(\mathbf{r}) e^{-i\omega_m t} \hat{b}_m + H.C. \right] . \tag{3.99}$$

These are quantum-field operators corresponding to (3.91) and (3.92). Hence, by using the commutation relations (3.95) and the completeness relation (3.76) we obtain

$$C_m = \left(\frac{\hbar}{2\omega_m \Omega_0} \right)^{1/2} . \tag{3.100}$$

We have thus canonically quantised the long wavelength polar optical oscillations, thus introducing the corresponding quantised longwave polar optical modes consistent with the general continuum model. The electron-phonon interaction Hamiltonian is then given by

$$\boxed{\hat{H}_F = \sum_m e \left(\frac{\hbar}{2\omega_m \Omega_0} \right)^{1/2} \left[\varphi_m(\mathbf{r}) e^{-i\omega_m t} \hat{b}_m + H.C. \right] .} \tag{3.101}$$

As explained above the function $\varphi_m(\mathbf{r})$ is obtained from a solution of the eigenvalue problem (3.57) which takes account of the coupling between electrical and mechanical excitations and accounts for all matching boundary conditions, mechanical and electrical, without *ad hoc* assumptions. Thus one has a firm basis for the study of the various problems of interest involving the interaction of electrons and long wave polar optical modes in semiconductor nanostructures.

The form of the amplitudes C_m (3.100) has long been known for acoustic modes in systems — e.g. a semiinfinite medium — with planar geometry [58–60]. The derivation given here holds quite generally (a) for any heterostructure with arbitrary geometry — e.g. cylindrical or spherical — and (b) for the polar optical modes, where $\mathbf{u}$ and φ are coupled and φ does not enter the orthogonality and completeness relations, which makes the case less obvious and requires explicit study.

Note that, as stressed in section 3, the eigenvectors $\mathbf{u}_m$ — and hence also the φ_m — contain an arbitrary normalisation constant $\Omega_0^{1/2}$, but the amplitude factor C_m of the $\mathbf{u}$ or φ field result from the application of a canonical rule. The arbitrary factor Ω_0 cancels out and so the actual field amplitudes, as well as $\hat{H}_F$ in (3.101) are uniquely determined.

Appendix A

In order to take into account the bulk phonon dispersion along certain directions in the Brillouin Zone we define a crystal free energy density associated with the tensor λ as

$$F = \frac{1}{2}\lambda_{iklm} u_{ik} u_{lm} \; ; \; (i, k, l, m = 1, 2, 3) , \qquad (3.102)$$

where $u_{ik} = \frac{1}{2}\left(\frac{\partial u_i}{\partial x_i} + \frac{\partial u_k}{\partial x_i}\right)$, with $x_1 = x, x_2 = y, x_3 = z$, plays the role of a 'strain tensor'. This term describes the energy related to the internal stress of the medium due to the optical vibrations. The tensor λ_{iklm} has symmetry properties leaving invariant the Lagrangian density (3.1), that is

$$\lambda_{iklm} = \lambda_{kilm} = \lambda_{ikml} = \lambda_{lmik} . \qquad (3.103)$$

Thus, out of the 81 components of λ_{iklm} only 21 are different. For cubic symmetry and taking rectangular coordinates as the principal axes of the crystal, we obtain reflection symmetry with respect to the three axes, that is the transformation $x_i \rightarrow -x_i$ ($i = 1, 2, 3$) must leave invariant the free energy. Hence, any component like $\lambda_{1112}, \lambda_{1222}, \lambda_{2333}$, etc. vanishes. For the cubic symmetry case all the coordinate axes are fourfold symmetry axes and we have three different nonzero components

$$\lambda_1 = \lambda_{1111} = \lambda_{2222} = \lambda_{3333}$$

$$\lambda_2 = \lambda_{1122} = \lambda_{1133} = \lambda_{2233} = \ldots \qquad (3.104)$$

$$\lambda_3 = \lambda_{1212} = \lambda_{2112} = \ldots$$

Throughout this book we take

$$\lambda_1 = 2\rho\beta_L^2, \quad \lambda_2 = 2\rho(\beta_L^2 - 2\beta_T^2), \quad \lambda_3 = \tfrac{\rho}{2}\beta_C^2 \qquad (3.105)$$

References

1. T. Ando, A. Fowler and F. Stern, Rev. Mod. Phys. **54**, 437 (1982).
2. L. Pfiffer, K.W. West, H.L. Stormer and K.W. Baldwin, Appl. Phys. Lett. **55**, 1888 (1989).
3. B. Vinter, Appl. Phys. Lett. **45**, 581 (1984).
4. L.T.P. Allen, E.R. Weber, J. Washburn and Y.C. Pao, Appl. Phys. Lett. **51**, 670 (1987).
5. Light Scattering in Solids V, edited by M. Cardona and G. Güntherodt, Springer, Berlin (1989).
6. Light Scattering is Solids VI, edited by M. Cardona and G. Güntherodt, Springer, Berlin (1993).
7. L. Esaki and R. Tsu, IBM J. Res. Dev. **14**, 61 (1970).
8. Z.V. Popović, M. Cardona, E. Richter, D. Strauch, L. Tapfer and K. Ploog, Phys. Rev. **B40**, 3040 (1989).
9. C. Trallero-Giner and F. Comas, Phys. Rev. **B37**, 4583 (1988).
10. M. Host, V. Norkt and J.P. Kottans, Phys. Rev. Lett. **50**, 754 (1983).
11. T. Englert, D.C. Tsui, J.C. Portal, J. Berens and A. Gossard, Solid State Commun. **44**, 1301 (1982).
12. M.A. Brammel, R.J. Nicholas, J.C. Portal, M. Rozeghi and M.A. Poisson, Physica B+C **177&118**, 753 (1983).
13. A.K. Sood, J. Menéndez, M. Cardona and K. Ploog, Phys. Rev. Lett. **54**, 2111 (1985); Phys. Rev. Lett. **54**, (1985).

14. C. Colvard, R. Fisher, T.A. Gant, M.V. Klein, R. Merlin, H. Morkov and A.C. Gossard, Superlattices and Microstructures **1**, 81 (1985).

15. B. Jusserand, D. Paquet and A. Reegreny, Superlattices and Microstructures **1**, 61 (1985).

16. A. Huber, T. Egeler, W. Ettmüller, H. Rothfritz, G. Trankle and G. Abstreiter, Superlattices and Microstructures **9**, 309 (1991).

17. R. Hessmer, A. Huber, T. Egeler, M. Haines, G. Tränkle, G. Weimann and G. Abstreiter, Phys. Rev. **B46**, 4071 (1992).

18. A. Fasolino, E. Molinari and J.C. Maan, Phys. Rev. **B39**, 3923 (1989).

19. B.V. Shanabrook, B.R. Bennet and R.J. Wagner, Phys. Rev. **B48**, 17172 (1993).

20. R. Schorer, G. Abstreiterr, S. de Gironcoli, E. Molinari, H. Kibbel and E. Kasper, Solid State Electronics. **37**, 757 (1994).

21. B. Foreman, Phys. Rev. **B52**, 12260 (1995).

22. E. Molinari, S. Baroni, P. Giannozzi and S. de Gironcoli, Phys. Rev. **B45**, 4280 (1992).

23. E. Molinari, A. Fasolino, K. Kunc, Superlattices and Microstructures **2**, 397 (1986).

24. E. Richter and D. Strauch, Solid State Commun. **64**, 867 (1987).

25. K. Huang and B. Zhu, Phys. Rev. **B38**, 13377 (1988).

26. S.F. Ren, Y.H. Chu and C.Y. Chang, Phys. Rev. **B37**, 8899 (1988).

27. H. Akera and T. Ando, Phys. Rev. **B40**, 2914 (1989).

28. H. Gerecke and F. Bechstedt, Phys. Stat. Sol. (b) **156**, 151 (1989).

29. H. Rücker, E. Molinari and P. Lugli, Phys. Rev. **B44**, 3463 (1991).

30. S.F. Ren and Y.C. Chang, Phys. Rev. **B43**, 11857 (1991).

31. B. Zhu, Phys. Rev. **B44**, 1926 (1991).

32. F. Rossi, L. Rota, C. Bungaro, P. Lugli and E. Molinari, Phys. Rev. **B47**, 1695 (1993).

33. E.P. Pokatilov and S, E. Beril, Phys. Stat. Sol. (b) **118**, 567 (1983).

34. V.M. Fomin and E.P. Pokatilov, Phys. Stat. Sol. (b) **123**, 69 (1985).
35. F. Bechstedt and R. Enderlein, Phys. Stat. Sol. (b) **131**, 53 (1985).
36. N. Mori and T. Ando, Phys. Rev. **B40**, 6175 (1989).
37. K. Nash, Phys. Rev. **B46**, 7723 (1992).
38. G. Weber. Phys. Rev. **B46**, 16171 (1992).
39. R. Lassning, Phys. Rev. **B30**, 7132 (1984).
40. M. Babiker, J. Phys. C: Solid State Physics **19**, 683 (1986).
41. L. Wendler, Phys. Stat. Sol. (b) **129**, 513 (1985).
42. L. Wendler and R. Haupt, Phys. Stat. Sol. (b) **143**, 487 (1987).
43. R. Cher, D.L. Lin and T.F. Geroge, Phys. Rev. **B41**, 1435 (1990).
44. F. Comas and C. Trallero-Giner, Physica B **192**, 394 (1993).
45. M. Cardona, Superlattices and Microstructures **7**, 183 (1990).
46. H. Rucker, E. Molinari and P. Lugli, Phys. Rev. **B44**, 3463 (1991).
47. B.K. Ridley, Phys. Rev. **B47**, 4592 (1993).
48. B.K. Ridley, Phys. Rev. **B44**, 9002 (1991).
49. H. Goldstein, *Classical Mechanics*, Addison-Wesley Publishing Company, London (1964).
50. L.D. Landau and E.M. Lifshitz, *Course of Theoretical Physics*, Vol. 8, 'Electrodynamics of Continuous Media', Pergamon Press, Oxford (1970).
51. M. Born and K. Huang *Dynamical Theory of Crystal Lattices*, Clarendon Press, Oxford (1988).
52. M.E. Mora, R. Pérez-Alvarez and Ch.B. Sommers, J. Physique **46**, 1021 (1985).
53. F. García-Moliner, R. Pérez-Alvarez, H. Rodríguez-Coppola and V.R. Velasco, J. Phys. A: Math. Gen. **23**, 1405 (1990).
54. C. Trallero-Giner, F. García-Moliner, V.R. Velasco and M. Cardona, Phys. Rev. **B45**, 11944 (1992).

55. F. García Moliner and V.R. Velasco, *Theory of Single and Multiple Interfaces: The Method of Surface Green Function Matching*. World Scientific, Singapore (1992).

56. L.D. Landau and E.M. Lifshitz, *Fluid Mechanics*, Pergamon Press, Oxford (1959).

57. M.P. Chamberlain, C. Trallero-Giner, and M. Cardona, Phys. Rev. **B50**, 1611 (1994).

58. H. Ezawa, Ann. Phys. **67**, 438 (1971).

59. P.S. King and F.W. Sheard, Proc. R. Soc. London Ser. A **230**, 175 (1970).

60. M.C. Oliveros and D.R. Tilley, Phys. Stat. Sol. (b) **119**, 675 (1983)

CHAPTER FOUR

Surface Green Function Matching

In Chapter 3 we have seen a general method of solution of the field equations for isotropic homogeneous media. Appropriate linear combinations of these can be formed to comply with the matching boundary conditions, as will be seen in the next chapter. In this method we write down the amplitudes — eigenfunctions — of the solutions.

An alternative way to perform a matching analysis can be carried out in terms of Green functions. The formalism is the same for either isotropic or anisotropic media and for interfaces of arbitrary geometry. Here we discuss it for isotropic media and for planar interfaces. One practical purpose for which Green functions are very useful is to obtain spectral functions very directly. This is done in this chapter to discuss in a technical way the hybrid nature of the normal polar optical modes in heterostructures.

While the standard matching boundary conditions are uniquely defined, some reasonable approximations can be sometimes made which amount to special matching conditions. For instance, a GaAs/AlAs interface, if we are interested in, say, the GaAs-like modes, it is an excellent approximation to assume that AlAs behaves as an infinitely rigid mechanical wall. The corresponding matching condition then takes a different form, but this applies only to the mechanical part of the waves. Full electrostatic matching is still required and so we are in a situation in which two coupled fields obey different types of matching boundary conditions. We discuss this issue and show how this is naturally incorporated in the Green function analysis of the matching problem for a general situation involving two coupled fields when one of them obeys a special type of matching — or surface — boundary conditions while the other one is subject to full matching rules.

We shall explicitly discuss the case in which the constituent media are piecewise homogeneous, but we recall from the general discussion of Chapter 3 that this method can also be used if inhomogeneities are involved.

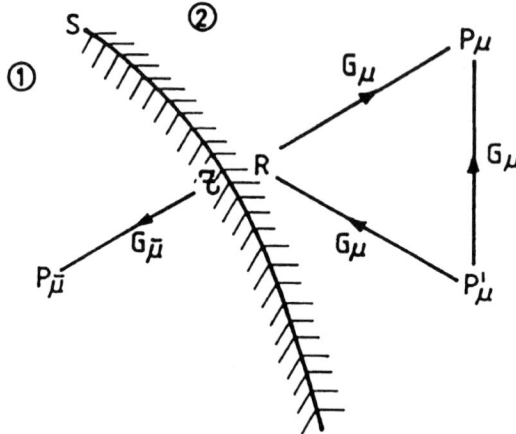

Figure 4.1 The propagator picture of the Surface Green Function Matching analysis. Propagation from a point in medium μ to another point in the same medium can proceed either directly or after reflection at the interface, while a point on the other side $\bar{\mu}$ can only be reached through transmission across the interface.

The Surface Green Function Matching method

The problems discussed in Chapter 3 can be alternatively formulated and solved in terms of Green functions. We shall now present a brief outline of the *Surface Green Function Matching* method which is fully discussed elsewhere [1].

Consider the matching problem depicted in Figure 4.1. Two media, labelled $\mu = 1, 2$, are matched at some interface, i.e., at some geometrical surface S. We assume we know the Green Functions G_μ of the (bulk) extended infinite media μ. An elementary excitation starts from some point P'_μ and reaches (i) another point P_μ in the same medium by direct propagation and by reflection $\mathcal{R}$ or (ii) a point $P_{\bar{\mu}}$ in the other medium — $\bar{\mu} = 2/1$ when $\mu = 1/2$ — after transmission $\mathcal{T}$. The problem is to find G_s, the Green Function of the matched system under study, given that we know G_1 and G_2.

The formal analysis is generally valid for any geometry of the matching surface [1, Chapter 6], but in view of the applications here contemplated we shall give the formulae for planar geometries. Then the objects appearing in the analysis are functions of z — or z and z' — of κ-2D wavevector in the (x, y) surface plane — and of the eigenvalue Ω. We now list the main results of the

Surface Green Function Matching analysis.

- The matrix elements of G_s can be written down exactly in terms of its surface projection $\mathcal{G}_s$ as

$$G_s(z,z') = \begin{cases} G_\mu(z,z') + G_\mu(z,0)\mathcal{G}_\mu^{-1}(\mathcal{G}_s - \mathcal{G}_\mu)\mathcal{G}_\mu^{-1}G_\mu(0,z'), & z,z' \in \mu \\ G_{\bar\mu}(z,0)\mathcal{G}_{\bar\mu}^{-1}\mathcal{G}_s\mathcal{G}_\mu^{-1}G_\mu(0,z'), & z \in \bar\mu, z' \in \mu. \end{cases} \quad (4.1)$$

The surface projections — functions only of (κ, Ω) — and their inverses, when they exist, are defined only within the 2D space of the surface. We note that G in general can be a matrix — in fact it *is* in our case — and then the products must be understood as contracted — scalar — products, so it is important to maintain the correct order of the factors everywhere.

- Consider an incident elementary excitation of medium μ, described by some amplitude $\mathbf{F}_\mu$. Then the scattering state solution for the same eigenvalue is

$$\mathbf{F}_s(z) = \begin{cases} \mathbf{F}_\mu(z) + G_\mu(z,0)\mathcal{G}_\mu^{-1}(\mathcal{G}_s - \mathcal{G}_\mu)\mathcal{G}_\mu^{-1}\mathbf{F}_{\mu 0}, & z \in \mu \\ G_{\bar\mu}(z,0)\mathcal{G}_{\bar\mu}^{-1}\mathcal{G}_s\mathcal{G}_\mu^{-1}\mathbf{F}_{\mu 0}, & z \in \bar\mu \end{cases}, \quad (4.2)$$

where $\mathbf{F}_{\mu 0}$ is the amplitude $\mathbf{F}_\mu$ at the surface $z = 0$.

- For Ω outside the range of allowed bulk eigenvalues a matched system may have distinct *matching solutions* given by the roots of the secular determinant

$$\det|\mathcal{G}_s^{-1}(\kappa,\Omega)| = 0. \quad (4.3)$$

- Then, if (4.3) has some zero at some $\Omega_s(\kappa)$, the amplitude of the matching state $\mathbf{F}_s$ — physically an interface state — for $\Omega = \Omega_s$ is

$$\mathbf{F}_s(z) = G_\mu(z,0)\mathcal{G}_\mu^{-1}\psi_s(0). \quad (4.4)$$

- Alternatively, from the full $G_s(\Omega,\kappa;z,z')$ we can obtain the various spectral functions of interest, namely $\mathcal{N}_s(\Omega,\kappa,z)$, $\mathcal{N}_s(\Omega,z)$ and $\mathcal{N}_s(\Omega)$, which are, respectively, the local κ resolved, the local and the total mode density. This point requires some clarification.

The Green function considered here — i.e. in this book, for polar optical modes — is by definition the resolvent of the linear differential operator

$$\hat{\mathbf{D}}_\kappa(z) = \rho\omega^2 \hat{\mathbf{I}} - \hat{\mathbf{L}}_\kappa(z), \tag{4.5}$$

where $\hat{\mathbf{L}}_\kappa$ is the differential operator introduced in Chapter 3 and $\hat{\mathbf{I}}$ is the 4×4 matrix

$$\begin{vmatrix} 1 & 0 & 0 & 0 \\ 0 & 1 & 0 & 0 \\ 0 & 0 & 1 & 0 \\ 0 & 0 & 0 & 0 \end{vmatrix}. \tag{4.6}$$

Correspondingly $\hat{\mathbf{L}}$ and $\hat{\mathbf{D}}$ are in general 4×4 differential matrices but in this Chapter we omit the explicit matrix notation.

From the matched $\mathbf{G}_s$ we obtain the spectral functions of interest starting from

$$\mathcal{N}(\Omega, \kappa, z) = -\frac{1}{\pi} \, Im \, Tr \, \mathbf{G}_s(\Omega, \kappa; z, z). \tag{4.7}$$

This gives the local κ resolved density of states and from here one can obtain any other spectral function of interest by integrating over κ and/or z. Note that in the formulation of Chapter 3, ρ is a weight factor -which, for a matched system, depends on z- and the eigenvalue Ω is ω^2.

It is customary to refer to the ω-dependence and write, for instance, $\mathbf{G}(\omega)$. However, one must always observe that the actual eigenvalue is ω^2, which cannot be ignored in (4.7). Even if one writes $\mathbf{G}(\Omega)$ as $\mathbf{G}(\omega)$, the actual mode densities *in* ω are given by the general expression

$$\mathcal{N}(\omega) = 2\omega \mathcal{N}(\Omega), \tag{4.8}$$

a fact which must be kept in mind.

We note that this provides an alternative to (4.3). Indeed, the eigenvalues given by the zeros of (4.3) are manifested in peaks of the spectral functions, e.g. of (4.7). In fact looking for peaks of $\mathcal{N}$ is in many cases more practical than looking for zeros of a determinant. Furthermore, the matching problem may have resonant solutions. A real Ω — or ω— solution which is forbidden on one side of the interface resonates with the allowed bulk continuum of the other side. Formally this results in a complex root of (4.3) but it is far more practical to study these situations by means of the spectral functions, as will be seen later.

The crucial question is the determination of $\mathbf{G}_s$, from which everything follows. The *matching formula* for $\mathbf{G}_s$ is derived by expressing the *matching* boundary conditions of Chapter 3 in Green function language. Note that this introduces normal derivatives of $\mathbf{G}(z, z')$ and these are discontinuous for $z' \to z \pm 0$, as follows easily from the inhomogeneous differential equation — or system — defining $\mathbf{G}$:

$$\hat{\mathbf{D}} \cdot \mathbf{G}(z, z') = \mathbf{I}\delta(z - z'), \tag{4.9}$$

with appropriate boundary conditions. Thus we must define the otherwise unspecified derivative. We define, at the surface $z = 0$,

$$'\mathcal{G}^{\pm} = \lim_{z' \to z \pm 0} \left[\frac{d\mathbf{G}(z, z')}{dz} \right]. \tag{4.10}$$

Now, the matching problem involves a primary field $\mathbf{F}(z)$, consisting physically of vibrational amplitudes and electrostatic potential, and a secondary field, consisting physically of electrical displacement $\mathbf{D}$ and 'mechanical stresses' σ_{ij}, all of which are linear differential forms in $\mathbf{F}$. The continuities of D_z and the σ_{iz} express the matching boundary conditions, which therefore involve the corresponding linear differential forms.

Now, just as the continuity of $\mathbf{F}$ entails the continuity of $\mathbf{G}$ for $z' \to z \pm 0$, so the continuity of the secondary field entails a discontinuity of the corresponding linear differential form in $\mathbf{G}$. The two normal derivatives of (4.10) originate two differential forms $\mathcal{A}^{(\pm)}$. The explicit formula for $\mathcal{A}^{(\pm)}$ depends on the specific problem under study, but these always have the fundamental discontinuity property

$$\mathcal{A}^{(+)} - \mathcal{A}^{(-)} = \mathbf{I} \tag{4.11}$$

and the combination of this with (4.2) — i.e. the actual matching analysis — yields the general matching formula

$$\boxed{\mathcal{G}_s^{-1} = \left(\mathcal{A}_1^{(+)} \mathcal{G}_1^{-1} \right) - \left(\mathcal{A}_2^{(-)} \mathcal{G}_2^{-1} \right),} \tag{4.12}$$

with the sign convention that medium 1 is in $z < 0$ and medium 2 in $z > 0$. All the details of this analysis can be found in Chapter 3 of Ref. [1] but it must be noted that the present sign convention for $\mathcal{A}^{(\pm)}$ is the opposite of Ref. [1].

We now summarise the basic elements needed for the Surface Green Function Matching analysis of the problem under study. Firstly we consider the full 3D Fourier transform of the linear differential matrix $\hat{\mathbf{D}}$ of a given bulk medium. This yields a 4×4 matrix $\mathbf{D}$ which is a function of Ω and $\mathbf{k} = (\kappa, k_z)$. We stress that the notation $\mathbf{D}$ is only a matter of expediency but it does not imply that this matrix is diagonal. In fact it usually is not. The matrix representing the Green function — the resolvent of $\hat{\mathbf{D}}$ — is the inverse

$$\mathbf{G}(\Omega, \mathbf{k}) = \mathbf{D}^{-1}(\Omega, \mathbf{k}), \qquad (4.13)$$

whence we obtain the (z, z')-dependent $\mathbf{G}$ in 2D Fourier transform

$$\mathbf{G}(\Omega, \kappa; z, z') = \int \frac{dk_z}{2\pi} \, \mathbf{G}(\Omega, \kappa, k_z) \, e^{ik_z(z-z')} \qquad (4.14)$$

from which we calculate normal derivatives and surface projections as needed for the surface Green function matching analysis [1].

Now, due to the isotropy of the model we have full rotational invariance in the (x, y) plane and thus we can chose the vector κ in the y-direction, $\kappa = (0, \kappa)$ without loss of generality. We then find that $\mathbf{D}(\Omega, \kappa)$ takes the form [2]

$$\mathbf{D} = \begin{vmatrix} D_{xx} & 0 & 0 & 0 \\ 0 & D_{yy} & D_{yz} & D_{y\varphi} \\ 0 & D_{zy} & D_{zz} & D_{z\varphi} \\ 0 & D_{\varphi y} & D_{\varphi z} & D_{\varphi\varphi} \end{vmatrix}. \qquad (4.15)$$

Consequently the G_{xx} component also factorises out, yielding

$$G_{xx} = D_{xx}^{-1} = \left[\rho \left(\omega^2 - \omega_T^2 \right) + \rho \beta_T^2 \left(\kappa^2 + k_z^2 \right) \right]^{-1}. \qquad (4.16)$$

This describes the usual type of *transverse horizontal* mode well known in the theory of elastic surface waves [3]. In fact the formula is the same but here it is written in terms of formal 'stiffness coefficients' and it is in order to observe (i) that the terms denoted 'stresses' are actually terms of the phenomenological model introduced in Chapter 3, which *have the nature* of stresses and their role is to introduce spatial dispersion in the model; and (ii) that the formal stiffness coefficients need not have a positive sign, as in the theory of elastic waves, where ω increases with increasing k. In fact in most cases — particularly in III-V compounds — ω decreases with increasing k, so the formal stiffness coefficients are negative.

This factorisation means that the transverse horizontal vibration u_x is decoupled from the rest. Our interest is in the other type of solution, for which we redefine $\mathbf{F}$ so, instead of being the four component column matrix of (3.56) it is now by new definition the three-component amplitude

$$\mathbf{F} = \begin{vmatrix} u_y \\ u_z \\ \varphi \end{vmatrix}, \qquad (4.17)$$

and consequently we redefine $\mathbf{D}(\Omega, \mathbf{k})$ and $\mathbf{G}(\Omega, \mathbf{k})$ as 3×3 matrices. For the purpose of the algebra we also relabel (y, z, φ) as $(1, 2, 3)$ respectively. Then the elements of the new $\mathbf{D}(\Omega, \mathbf{k})$ matrix are

$$\begin{aligned}
D_{11} &= \rho(\omega^2 - \omega_T^2) + \rho \beta_T^2 k_z^2 + \rho \beta_L^2 \kappa^2 \\
D_{12} &= \rho(\beta_L^2 - \beta_T^2) \kappa k_z \\
D_{13} &= -i\alpha\kappa \\
D_{21} &= \rho(\beta_L^2 - \beta_T^2) \kappa k_z \\
D_{22} &= \rho(\omega^2 - \omega_T^2) + \rho \beta_T^2 \kappa^2 + \rho \beta_L^2 k_z^2 \\
D_{23} &= -i\alpha k_z \\
D_{31} &= \frac{4\pi}{\epsilon_\infty} i\alpha\kappa \\
D_{32} &= \frac{4\pi}{\epsilon_\infty} i\alpha k_z \\
D_{33} &= (\kappa^2 + k_z^2) .
\end{aligned} \qquad (4.18)$$

In the following we denote

$$\Delta = \rho^2 \beta_L^2 \beta_T^2 (\kappa^2 + k_z^2)(k_z^2 - k_T^2)(k_z^2 - k_L^2), \qquad (4.19)$$

$$k_L^2 = \frac{\omega_L^2 - \omega^2}{\beta_L^2} - \kappa^2, \qquad (4.20)$$

$$k_T^2 = \frac{\omega_T^2 - \omega^2}{\beta_T^2} - \kappa^2. \qquad (4.21)$$

Then the elements of the inverse (4.13) are

$$G_{11}(\kappa, k_z) = \left[-\frac{k_T^2 \kappa^2}{\beta_L^2} - \left(\frac{k_L^2}{\beta_T^2} - \frac{\kappa^2}{\beta_L^2} \right) k_z^2 + \frac{k_z^4}{\beta_T^2} \right] \times \left[\rho(\kappa^2 + k_z^2)(k_z^2 - k_T^2)(k_z^2 - k_L^2) \right]^{-1}$$

$$G_{12}(\kappa, k_z) = \frac{\left[\left(\frac{1}{\beta_L^2} - \frac{1}{\beta_T^2} \right) \kappa^3 + \frac{(\omega_L^2 - \omega_T^2)}{\beta_T^2 \beta_L^2} \kappa \right] k_z + \left(\frac{1}{\beta_L^2} - \frac{1}{\beta_T^2} \right) \kappa k_z^3}{\rho(\kappa^2 + k_z^2)(k_z^2 - k_T^2)(k_z^2 - k_L^2)}$$

$$G_{13}(\kappa, k_z) = \frac{i\alpha\kappa}{\rho \beta_L^2 (\kappa^2 + k_z^2)(k_z^2 - k_L^2)}$$

$$G_{21}(\kappa, k_z) = G_{12}(\kappa, k_z)$$

$$G_{22}(\kappa, k_z) = \left[-\frac{k_T^2 k_z^2}{\beta_L^2} - \left(\frac{k_L^2}{\beta_T^2} - \frac{k_z^2}{\beta_T^2}\right)\kappa^2 + \frac{k_z^4}{\beta_L^2}\right] \times$$
$$\left[\rho(\kappa^2 + k_z^2)(k_z^2 - k_T^2)(k_z^2 - k_L^2)\right]^{-1}$$

$$G_{23}(\kappa, k_z) = \frac{i\alpha k_z}{\rho\beta_L^2(\kappa^2 + k_z^2)(k_z^2 - k_L^2)}$$

$$G_{31}(\kappa, k_z) = -\left(\frac{4\pi}{\epsilon_\infty}\right) G_{13}(\kappa, k_z)$$

$$G_{32}(\kappa, k_z) = -\left(\frac{4\pi}{\epsilon_\infty}\right) G_{23}(\kappa, k_z)$$

$$G_{33}(\kappa, k_z) = \frac{1}{k_z^2 - k_L^2} - \frac{\beta_T^2}{\beta_L^2}\frac{(k_T^2 + \kappa^2)}{(k_T^2 + \kappa^2)(k_z^2 - k_L^2)}. \tag{4.22}$$

By performing the 1D Fourier transform of (4.14) this yields the $G_{ij}(\Omega, \kappa; z, z')$:

$$G_{11}(\kappa; z, z') = \frac{\kappa}{2}\frac{\left[\frac{k_L^2}{\beta_T^2} - \frac{k_T^2}{\beta_L^2} + \kappa^2\left(\frac{1}{\beta_T^2} - \frac{1}{\beta_L^2}\right)\right]}{\rho(k_T^2 + \kappa^2)(k_L^2 + \kappa^2)} e^{-\kappa|z-z'|}$$
$$+ \frac{i}{2}\frac{k_T}{\rho\beta_T^2(k_T^2 + \kappa^2)} e^{ik_T|z-z'|}$$
$$+ \frac{i}{2}\frac{\kappa^2}{\rho\beta_L^2 k_L(k_L^2 + \kappa^2)} e^{ik_L|z-z'|}$$

$$G_{12}(\kappa; z, z') = \frac{i}{2}\text{sgn}(z-z')\left[\frac{\kappa(\omega_L^2 - \omega_T^2)}{\rho\beta_L^2\beta_T^2(k_T^2 + \kappa^2)(k_L^2 + \kappa^2)} e^{-\kappa|z-z'|}\right.$$
$$\left.+ \frac{\left[\left(\frac{1}{\beta_L^2} - \frac{1}{\beta_T^2}\right)\kappa(k_T^2 + \kappa^2) + \kappa\frac{(\omega_L^2 - \omega_T^2)}{\beta_L^2\beta_T^2}\right]}{\rho(k_T^2 + \kappa^2)(k_T^2 - k_L^2)} e^{ik_T|z-z'|}\right.$$

$$+ \frac{\left[\left(\frac{1}{\beta_L^2} - \frac{1}{\beta_T^2}\right)\kappa(k_L^2 + \kappa^2) + \frac{(\omega_L^2 - \omega_T^2)}{\beta_L^2 \beta_T^2}\kappa\right]}{\rho(k_L^2 + \kappa^2)(k_L^2 - k_T^2)} e^{ik_L|z-z'|}\Bigg]$$

$$G_{13}(\kappa; z, z') = \frac{i\alpha\kappa}{2\rho\beta_L^2}\left[-\frac{e^{-\kappa|z-z'|}}{\kappa(k_L^2 + \kappa^2)} + \frac{i\,e^{ik_L|z-z'|}}{k_L(k_L^2 + \kappa^2)}\right]$$

$$G_{21}(\kappa; z, z') = G_{12}(\kappa; z, z')$$

$$G_{22}(\kappa; z, z') = \frac{\kappa}{2}\frac{\left[\frac{k_T^2}{\beta_L^2} - \frac{k_L^2}{\beta_T^2} + \left(\frac{1}{\beta_L^2} - \frac{1}{\beta_T^2}\right)\kappa^2\right]}{\rho(k_T^2 + \kappa^2)(k_L^2 + \kappa^2)} e^{-\kappa|z-z'|}$$

$$+ \frac{i}{2}\frac{\kappa^2}{\rho\beta_T^2 k_T(k_T^2 + \kappa^2)} e^{ik_T|z-z'|}$$

$$+ \frac{i}{2}\frac{k_L}{\rho\beta_L^2(k_L^2 + \kappa^2)} e^{ik_L|z-z'|}$$

$$G_{23}(\kappa; z, z') = \frac{\alpha}{2\rho\beta_L^2}sgn(z - z')\left[\frac{e^{-\kappa|z-z'|}}{k_L^2 + \kappa^2} - \frac{e^{ik_L|z-z'|}}{k_L^2 + \kappa^2}\right]$$

$$G_{31}(\kappa; z, z') = -\left(\frac{4\pi}{\epsilon_\infty}\right)G_{13}(\kappa; z, z')$$

$$G_{32}(\kappa; z, z') = -\left(\frac{4\pi}{\epsilon_\infty}\right)G_{23}(\kappa; z, z')$$

$$G_{33}(\kappa; z, z') = \frac{\beta_T^2}{\beta_L^2}\frac{(k_T^2 + \kappa^2)}{2\kappa(k_L^2 + \kappa^2)} e^{-\kappa|z-z'|}$$

$$+ \frac{i}{2k_L}\left(1 - \frac{\beta_T^2}{\beta_L^2}\frac{(k_T^2 + \kappa^2)}{(k_L^2 + \kappa^2)}\right) e^{ik_L|z-z'|} \, . \qquad (4.23)$$

These elements depend on $(z - z')$ because so far we are looking at a bulk Green function; the final $\mathbf{G}_s$ of course will depend on z and z' separately. The surface projection of (4.23) yields

$$\mathcal{G}_{11}(\kappa) = \frac{1}{2\rho\beta_T^2(\kappa - ik_T)} - \frac{\kappa}{2\rho\beta_L^2 k_L(k_L + i\kappa)}$$

$$\mathcal{G}_{12}(\kappa) = 0$$

$$\mathcal{G}_{13}(\kappa) = -\frac{\alpha}{2\rho\beta_L^2 k_L(\kappa - ik_L)}$$

$$\mathcal{G}_{21}(\kappa) = 0$$

$$\mathcal{G}_{22}(\kappa) = \frac{1}{2\rho\beta_L^2(\kappa - ik_L)} - \frac{\kappa}{2\rho\beta_T^2 k_T(k_T + i\kappa)}$$

$$\mathcal{G}_{23}(\kappa) = 0$$

$$\mathcal{G}_{31}(\kappa) = -\left(\frac{4\pi}{\epsilon_\infty}\right)\mathcal{G}_{13}(\kappa)$$

$$\mathcal{G}_{32}(\kappa) = 0$$

$$\mathcal{G}_{33}(\kappa) = \frac{\beta_T^2(k_T^2 + \kappa^2) + \kappa\beta_L^2(ik_L - \kappa)}{2\beta_L^2 \kappa k_L(k_L + i\kappa)}. \qquad (4.24)$$

The derivatives of (4.24) are readily obtained and then, after (4.10), the elements of $'\mathcal{G}^{(\pm)}$ are

$${'\mathcal{G}}_{11}^{\pm}(\kappa) = \pm\frac{1}{2\rho\beta_T^2}$$

$${'\mathcal{G}}_{12}^{\pm}(\kappa) = \frac{\kappa(\beta_L^2 - \beta_T^2)(k_L + i\kappa)(k_T + i\kappa) + \kappa(\omega_L^2 - \omega_T^2)}{2\rho\beta_L^2\beta_T^2(k_T + k_L)(k_L + i\kappa)(k_T + i\kappa)}$$

$${'\mathcal{G}}_{13}^{\pm}(\kappa) = 0$$

$${'\mathcal{G}}_{21}^{\pm}(\kappa) = {'\mathcal{G}}_{12}^{\pm}(\kappa)$$

$${'\mathcal{G}}_{22}^{\pm}(\kappa) = \pm\frac{1}{2\rho\beta_L^2}$$

$${'\mathcal{G}}_{23}^{\pm}(\kappa) = -\frac{\alpha}{2\rho\beta_L^2(\kappa - ik_L)}$$

$${'\mathcal{G}}_{31}^{\pm}(\kappa) = 0$$

$${'\mathcal{G}}_{32}^{\pm}(\kappa) = -\left(\frac{4\pi}{\epsilon_\infty}\right){'\mathcal{G}}_{23}^{\pm}(\kappa)$$

$${'\mathcal{G}}_{33}^{\pm}(\kappa) = \pm\frac{1}{2}. \qquad (4.25)$$

Again, for a bulk homogeneous case these elements do not depend on the point taken as $z = 0$, but this will not be the case for the matched system.

Finally, in this model the linear differential form $\mathcal{A}^{(\pm)}$, which results from the form of the tensor σ with $\mathbf{u}$ replaced by $\mathbf{G}$, is

$$\mathcal{A}_{11}^{\pm}(\kappa) = \rho\beta_T^2 \left({'\mathcal{G}}_{11}^{\pm} + i\kappa\mathcal{G}_{21}\right)$$

$$\begin{aligned}
\mathcal{A}_{12}^{\pm}(\kappa) &= \rho\beta_T^2\left('\mathcal{G}_{12}^{\pm} + i\kappa\mathcal{G}_{22}\right) \\
\mathcal{A}_{13}^{\pm}(\kappa) &= \rho\beta_T^2\left('\mathcal{G}_{13}^{\pm} + i\kappa\mathcal{G}_{23}\right) \\
\mathcal{A}_{21}^{\pm}(\kappa) &= \rho(\beta_L^2 - 2\beta_T^2)i\kappa\mathcal{G}_{11} + \rho\beta_L^2\,'\mathcal{G}_{21}^{\pm} \\
\mathcal{A}_{22}^{\pm}(\kappa) &= \rho(\beta_L^2 - 2\beta_T^2)i\kappa\mathcal{G}_{12} + \rho\beta_L^2\,'\mathcal{G}_{22}^{\pm} \\
\mathcal{A}_{23}^{\pm}(\kappa) &= \rho(\beta_L^2 - 2\beta_T^2)i\kappa\mathcal{G}_{13} + \rho\beta_L^2\,'\mathcal{G}_{23}^{\pm} \\
\mathcal{A}_{31}^{\pm}(\kappa) &= -\left(\frac{4\pi\alpha}{\epsilon_\infty}\right)\mathcal{G}_{21} + '\mathcal{G}_{31}^{\pm} \\
\mathcal{A}_{32}^{\pm}(\kappa) &= -\left(\frac{4\pi\alpha}{\epsilon_\infty}\right)\mathcal{G}_{22} + '\mathcal{G}_{32}^{\pm} \\
\mathcal{A}_{33}^{\pm}(\kappa) &= -\left(\frac{4\pi\alpha}{\epsilon_\infty}\right)\mathcal{G}_{23} + '\mathcal{G}_{33}^{\pm}. \quad (4.26)
\end{aligned}$$

With these results we can now perform the Surface Green Function Matching analysis of the matched system.

Planar structures

Most heterostructures of physical interest involve more than one interface. From the point of view of any matching analysis this means that matching must be simultaneously performed at N independent interfaces. In the case of Surface Green Function Matching this requires an appropriate extension of the method outlined in section 4, which we shall now consider for various planar heterostructures of interest.

A sandwich structure of the type 1–2–3

This could be a general asymmetric quantum well, or a symmetric one with media 1 and 3 equal. However, for the purpose of the matching analysis it is convenient to retain the labels 1 and 3 denoting different geometrical domains. The details are given in Ref. [1] (Chapter 5).

Now, the form of the matched $\mathbf{G}_s$ for the single interface (4.2) has an interesting meaning in terms of scattering theory. Consider the case in which z and z' are on the same side of the interface. An elementary excitation of the medium in question sees the interface as a perturbation of the homogeneous bulk which causes scattering, so we can view $\mathbf{G}_s$ as the scattered propagator. But we note that (4.2) gives $\mathbf{G}_s$ exactly in terms of $\mathcal{G}_s$, for which there is a formula. Thus the Surface Green Function Matching analysis amounts to transforming the scattering problem into a matching problem and then obtaining the exact t-matrix for

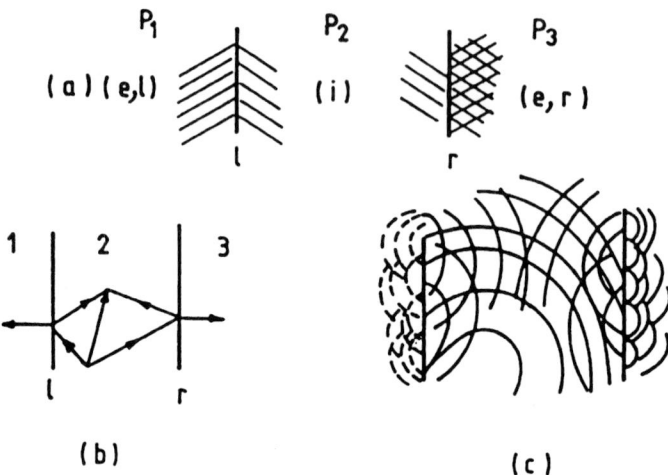

Figure 4.2 Schematic view of a layered structure of the 'sandwich' type. (a): The three domains of the structure with corresponding domains. The external domain consists of P_1 on the left and P_3 on the right. The two media can be physically the same material, but the geometrical label still distinguishes the two disconnected subdomains of the external domain. The internal domain is P_2. (b): Ray picture. (c): Wave picture.

reflection scattering. The theory also accounts for the other possibility of *transmission* scattering, when z and z' are on opposite sides of the interface. The point is to note that in all cases obtaining the corresponding **t**-matrix amounts to solving the scattering problem to infinite order in the perturbation [1] and this is particularly useful when it comes to treating structures with more than one interface.

The situation for the 1-2-3 sandwich is illustrated in Figure 4.2: We have the internal domain (i) and the external domain consisting of the $(e, l)/(e, r)$ subdomain on the left/right. Consider an elementary excitation starting from inside: (b) gives the ray picture and (c) the wave picture. The reflected amplitude after reflection at, say, the l interface could reach another interior point either directly or, again, indirectly after reflection at the r interface.

This would involve the summation of the series corresponding to infinitely many repeated reflections, but this is circumvented by defining *one* interface consisting of the two sheets l and r, consistently with the definition of just the internal and the external domains i and e, although the latter consists of two disconnected subdomains. We define the full unit $\mathcal{I}$ of the full matching surface as

$$\mathcal{I} = \mathcal{I}_l + \mathcal{I}_r . \tag{4.27}$$

Here $\mathcal{I}_l/\mathcal{I}_r$ denotes the surface projector on the left/right surface. Let P_μ be the projector on domain $\mu = 1, 2, 3$, with propagator $\mathbf{G}_\mu$. We define the *external* and *internal* domains with projectors

$$P_e = P_1 + P_3 \quad ; \quad P_i = P_2 , \tag{4.28}$$

and the external and internal propagators

$$\mathbf{G}_e = P_e \mathbf{G}_e P'_e = P_1 \mathbf{G}_1 P'_1 + P_3 \mathbf{G}_3 P'_3 ,$$
$$\mathbf{G}_i = P_i \mathbf{G}_i P'_i = P_2 \mathbf{G}_2 P'_2 . \tag{4.29}$$

We then define the full surface projections

$$\tilde{\mathbf{G}}_i = \tilde{\mathbf{G}}_2 = \begin{vmatrix} \mathcal{G}_{2l} & \mathbf{G}_2(l,r) \\ \mathbf{G}_2(r,l) & \mathcal{G}_{2r} \end{vmatrix} \quad ; \quad \tilde{\mathbf{G}}_e = \begin{vmatrix} \mathcal{G}_{1l} & 0 \\ 0 & \mathcal{G}_{3r} \end{vmatrix} \tag{4.30}$$

where $\mathcal{G}_{\mu l}/\mathcal{G}_{\mu r}$ is the former single surface projection of $\mathbf{G}_\mu$ at the l/r surface. In general each $\mathcal{G}_\mu$ is an n×n matrix and $\tilde{\mathbf{G}}$ is a 2n×2n matrix, which we shall term a 2×2 supermatrix. Likewise if the amplitude $\mathbf{F}$, and therefore also its single surface projection $\mathcal{F}_l/\mathcal{F}_r$ has n components, then its full interface projection

$$\mathcal{F}_s = \begin{vmatrix} \mathcal{F}_l \\ \mathcal{F}_r \end{vmatrix} \tag{4.31}$$

has $2n$ components. It is then proved [1, Chapter 5] that the Surface Green Function Matching algebras of the single interface and the sandwich problems are isomorphic with the *one* side and the *other* side in correspondence with the *in*side and *out*side and with quantities like $\mathcal{G}, \mathcal{F}$ replaced by $\tilde{\mathbf{G}}, \tilde{\mathcal{F}}$.

Let z_μ indicate that z is in subdomain $\mu = 1, 2, 3$. Then the form of $G_s(z_\mu, z'_\mu)$ depends on the (z_μ, z'_μ) configuration. We now give the formulae for the five distinct configurations:

$$\mathbf{G}_s(z_2, z'_2) = \mathbf{G}_2(z_2, z'_2) +$$
$$[\mathbf{G}_2(z_2, l), \mathbf{G}_2(z_2, r)] \cdot \tilde{\mathbf{G}}_2^{-1} \cdot \left(\tilde{\mathbf{G}}_s - \tilde{\mathbf{G}}_2 \right) \cdot \tilde{\mathbf{G}}_2^{-1} \cdot \begin{vmatrix} \mathbf{G}_2(l, z'_2) \\ \mathbf{G}_2(r, z'_2) \end{vmatrix} ,$$

$$\mathbf{G}_s(z_3, z'_2) = [0, \mathbf{G}_3(z_3, r)] \cdot \tilde{\mathbf{G}}_e^{-1} \cdot \tilde{\mathbf{G}}_s \cdot \tilde{\mathbf{G}}_2^{-1} \cdot \begin{vmatrix} \mathbf{G}_2(l, z'_2) \\ \mathbf{G}_2(r, z'_2) \end{vmatrix} ,$$

$$\mathbf{G}_s(z_2, z_3') = [\mathbf{G}_2(z_2, l), \mathbf{G}_2(z_2, r)] \cdot \tilde{\mathbf{G}}_2^{-1} \cdot \tilde{\mathbf{G}}_s \cdot \tilde{\mathbf{G}}_e^{-1} \cdot \begin{vmatrix} 0 \\ \mathbf{G}_3(r, z_3') \end{vmatrix},$$

$$\mathbf{G}_s(z_3, z_3') = \mathbf{G}_3(z_3, z_3') +$$
$$[0, \mathbf{G}_3(z_3, r)] \cdot \tilde{\mathbf{G}}_e^{-1} \cdot \left(\tilde{\mathbf{G}}_s - \tilde{\mathbf{G}}_e \right) \cdot \tilde{\mathbf{G}}_e^{-1} \cdot \begin{vmatrix} 0 \\ \mathbf{G}_3(r, z_3') \end{vmatrix},$$

$$\mathbf{G}_s(z_1, z_3') = [\mathbf{G}_1(z_1, l), 0] \cdot \tilde{\mathbf{G}}_e^{-1} \cdot \tilde{\mathbf{G}}_s \cdot \tilde{\mathbf{G}}_e^{-1} \begin{vmatrix} 0 \\ \mathbf{G}_3(r, z_3') \end{vmatrix}. \quad (4.32)$$

Note the appearance of *row supermatrices* — e.g. the first factor on the r.h.s. of $\mathbf{G}_s(z_2, z_2')$ — and *column supermatrices* — e.g. the last factor in the same term. The formulae for scattering state or matching state amplitudes can likewise be extended in an obvious way. In practice we shall be interested in the matching solutions [1, section 5.1].

$$\mathbf{F}_s(z_1) = \mathbf{G}_1(z_1, l) \cdot \mathcal{G}_{sl}^{-1} \cdot \mathcal{F}_l ,$$

$$\mathbf{F}_s(z_2) = [\mathbf{G}_2(z_2, l), \mathbf{G}_2(z_2, r)] \cdot \tilde{\mathbf{G}}_2^{-1} \cdot \tilde{\mathcal{F}}_s ,$$

$$\mathbf{F}_s(z_3) = \mathbf{G}(z_3, r) \cdot \mathcal{G}_{3r}^{-1} \cdot \mathcal{F}_r . \quad (4.33)$$

The next step is to extend the definition of the extended normal derivatives, for which we note that with the sign convention usually taken z grows from left to right across the system. We define

$$'\tilde{\mathbf{G}}_e = \begin{vmatrix} '\mathcal{G}_{1l}^{(+)} & 0 \\ 0 & -\mathcal{G}_{3r}^{(-)} \end{vmatrix},$$

$$'\tilde{\mathbf{G}}_i = \begin{vmatrix} '\mathcal{G}_{2l}^{(-)} & '\mathbf{G}_2(l, r) \\ -'\mathbf{G}_2(r, l) & -'\mathcal{G}_{2r}^{(+)} \end{vmatrix}. \quad (4.34)$$

The meaning of a term like $'\mathbf{G}_2(l, r)$ should be obvious: We differentiate with respect to z and then take z to l and z' to r while z and z' stay always within medium 2. By the same formulae of (4.26) projected at the successive interfaces, we obtain the corresponding 2×2 differential supermatrices $\tilde{\mathcal{A}}_e$ and $\tilde{\mathcal{A}}_i$ and then the matching formula is

$$\tilde{\mathbf{G}}_s^{-1} = \tilde{\mathcal{A}}_e \cdot \tilde{\mathbf{G}}_e^{-1} - \tilde{\mathcal{A}}_i \cdot \tilde{\mathbf{G}}_i^{-1} . \quad (4.35)$$

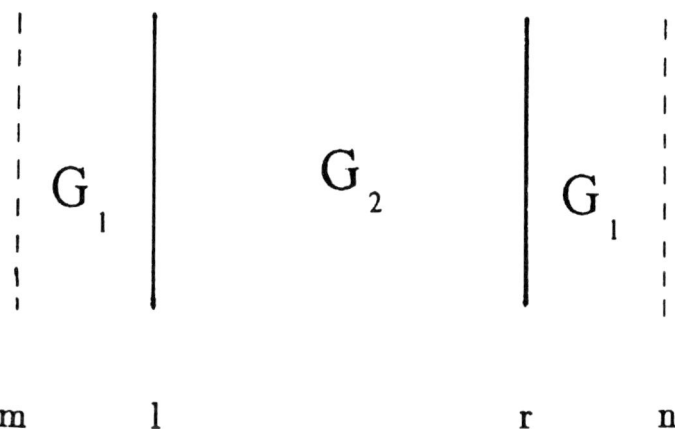

Figure 4.3 Schematic view of a superlattice. As in the sandwich structure we still define an internal domain — now optional — and an external one consisting of two disconnected — and now finite — subdomains.

We stress again that the sign convention for all $\mathcal{A}^{(\pm)}$ is here opposite to that of Ref. [1]. The isomorphism between the one interface and two interface cases is now complete.

The superlattice ...1–2–1–2–1–2–1–2–1–2...

As in the sandwich or quantum well case there are only two physically distinct interfaces; the rest is a periodic repetition. We define d_μ ($\mu = 1, 2$) as the thickness of the slab of medium μ. Then the superlattice period d is $d_1 + d_2$ and the Bloch phase factor is

$$f = e^{iqd}, \qquad (4.36)$$

where q is the 1D super wavevector associated with the superperiodicity. By convention we formally define a slab of medium 2 as the internal domain, while the pair of adjacent slabs constitute the external domain — Figure 4.3 — so (4.28) holds again, but now P_1 and P_3 are projectors of finite domains. The definitions of projections pertaining to the internal medium — $\tilde{G}_i, {}'\tilde{G}_i, \tilde{\mathcal{A}}_i$ — are as before. The difference appears with the quantities pertaining to the external medium, for which we define

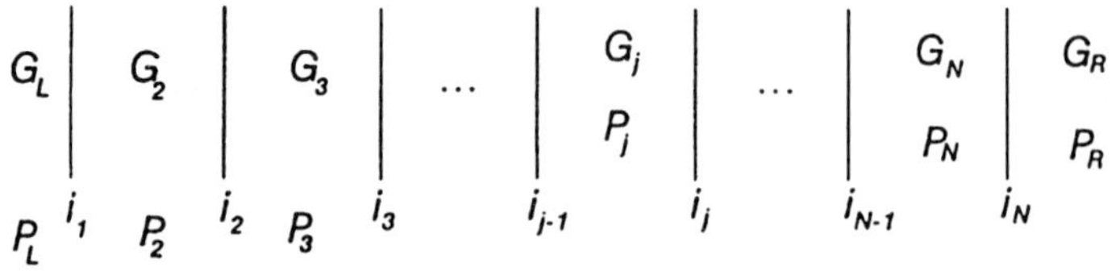

```
G_L | G_2 | G_3 | ... | G_j | ... | G_N | G_R
                       P_j           P_N   P_R
P_L  i_1  P_2  i_2  P_3  i_3  ...  i_{j-1}  i_j  ...  i_{N-1}  i_N
```

Figure 4.4 A general heterostructure with N interfaces and N domains. Note that domain 1 is disjoint and consists of L on the left and R on the right. The symbols i_j denote the projections on the different interfaces.

and

$$\hat{\mathbf{G}}_e = \begin{vmatrix} \mathcal{G}_{1l} & f^{-1}\mathbf{G}_1(l,m) \\ f\mathbf{G}_3(r,n) & \mathcal{G}_{3r} \end{vmatrix} \tag{4.37}$$

$$'\hat{\mathbf{G}}_e = \begin{vmatrix} '\mathcal{G}_{1l} & f^{-1} \,'\mathbf{G}_1(l,m) \\ -f\,'\mathbf{G}_1(r,n) & -'\mathcal{G}_{1r} \end{vmatrix}, \tag{4.38}$$

with the obvious implications on the corresponding $\hat{\mathcal{A}}_e$.

The matching formula is then like (4.35), namely

$$\tilde{\mathbf{G}}_s^{-1} = \hat{\mathcal{A}}_e \cdot \hat{\mathbf{G}}_e^{-1} - \hat{\mathcal{A}}_i \cdot \hat{\mathbf{G}}_i^{-1}, \tag{4.39}$$

with the only difference that $\tilde{\mathbf{G}}_e$, $\tilde{\mathcal{A}}_e$ are replaced by $\hat{\mathbf{G}}_e$, $\hat{\mathcal{A}}_e$, otherwise the isomorphism carries over.

The rest of the analysis is likewise extended (details in Ref. [1], section 5.1).

Structures with an arbitrary number of interfaces

This could be any multilayer structure, like for instance a finite multiple quantum well, a *Fibonacci* sequence or a tunnelling structure [4]. The notation is laid out in Figure 4.4: Domain 1 consists of the two disconnected subdomains L (left) and R (right) and i_j is the projector onto the j-th surface. Consider domain j, bounded by the interfaces with projectors i_{j-1} and i_j. We define the partial two-surface projector

$$\mathcal{I}_j = i_{j-1} + i_j. \tag{4.40}$$

Now, the extension of the formal algebra from one to two interfaces is effected by defining the simultaneous projection at the two corresponding surfaces

Surface Green Function Matching

and by going over from single surface projection $\mathcal{G}$ to two-surface projections $\tilde{\mathbf{G}}$ and likewise for the normal derivatives. The algebra is then isomorphic. The same process can be carried over to any arbitrary number N of the interfaces [5] by defining the simultaneous N-surface projections. We recall that a single surface projection, being a 2×2 supermatrix, is a $2n \times 2n$ matrix. For an arbitrary N, the N-surface projections are $N \times N$ supermatrices having tridiagonal form since each $\mathbf{G}_j$, for $j > 1$, has only the four submatrices resulting from the projection on the two adjacent interfaces, i.e.

$$\tilde{\mathbf{G}}_j = \left\| \begin{array}{cc} <i_{j-1}|\mathbf{G}_j|i_{j-1}> & <i_{j-1}|\mathbf{G}_j|i_j> \\ <i_j|\mathbf{G}_j|i_{j-1}> & <i_j|\mathbf{G}_j|i_j> \end{array} \right\| \equiv \mathcal{I}_j \tilde{\mathbf{G}}_j \mathcal{I}_j \quad (4.41)$$

for $j > 1$ and

$$\tilde{\mathbf{G}}_1 = \left\| \begin{array}{cc} <i_1|\mathbf{G}_L|i_1> & 0 \\ 0 & <i_N|\mathbf{G}_R|i_N> \end{array} \right\| \equiv \mathcal{I}_1 \tilde{\mathbf{G}}_1 \mathcal{I}_1 . \quad (4.42)$$

This is a practical feature when it comes to inverting the full $N \times N$ supermatrix, which, in fact needs to be done only once [1].

The first part of the Surface Green Function Matching analysis is then isomorphically extended in a straightforward way and yields the general results [5]

$$<z_j|\mathbf{G}_s|z'_j> = <z_j|\mathbf{G}_j|z'_j> +$$
$$<z_j|\mathbf{G}_s|\mathcal{I}'_j> \cdot \tilde{\mathbf{G}}_j^{-1} \cdot (\tilde{\mathbf{G}}_s - \tilde{\mathbf{G}}_j) \cdot \tilde{\mathbf{G}}_j^{-1} \cdot <\mathcal{I}_j|\mathbf{G}_j|z'_j>, \quad (4.43)$$

for z and z' in domain j and

$$<\mathcal{I}_j|\tilde{\mathbf{G}}_s|\mathcal{I}_k> = <\mathcal{I}_j|\tilde{\mathbf{G}}_j|\mathcal{I}_j> \cdot <\mathcal{I}_j|\mathcal{T}(j,k)|\mathcal{I}_k> \cdot <\mathcal{I}_k|\tilde{\mathbf{G}}_k|\mathcal{I}_k>, \quad (4.44)$$

for z in domain j and z' in domain $k \neq j$.

In order to obtain the matching formula for the full $\tilde{\mathbf{G}}_s^{-1}$ we need only express the same matching boundary conditions at all interfaces, for which we define for each domain the corresponding $'\tilde{\mathbf{G}}_j$ and $\tilde{\mathcal{A}}_j$ by an obvious extension of (4.38), which yields the result

$$\tilde{\mathbf{G}}_s^{-1} = \mathcal{I} \left(\tilde{\mathcal{A}}_1 \cdot \tilde{\mathbf{G}}_1^{-1} - \sum_{j=2}^{N} \tilde{\mathcal{A}}_j \cdot \tilde{\mathbf{G}}_j^{-1} \right) \mathcal{I}, \quad (4.45)$$

where $\mathcal{I}$ is the full N-surface projector

$$\mathcal{I} = \sum_{j=1}^{N} i_j . \qquad (4.46)$$

We note some practical aspects of the above results. Firstly, most of the algebra — i.e. the matrix inversion and multiplication — is carried out in the reduced subspace of the 2×2 supermatrices. *After* these operations have been carried out, the partial results are all put in the large N×N supermatrix format, where only the *four* — or *two*, for $j = 1$ — submatrices of each j-th term are non vanishing, and then added up. One then has to invert a tridiagonal supermatrix *only once*, to obtain $\tilde{\mathbf{G}}_s$ after (4.45) has been evaluated. Furthermore, to invert very large tridiagonal matrices one can use an algorithm which has been developed to perform another type of Surface Green Function Matching calculation [6] and which works very efficiently.

Systems with mirror-image symmetry

Consider, for instance, the double barrier structure A-B-A-B-A often used for resonant tunnelling experiments [4]. This is a four-interface problem, but the mirror image symmetry of the structure can be used to factorise two independent problems in orthogonal subspaces of even and odd solutions [7].

There is an important technical point to be considered, which often arises in practice and does arise in fact in the case we are concerned with, that is, the three-component amplitude $\mathbf{F}_\kappa(z)$ defined in (3.56). We stress that we have assumed we have factorised out the transverse horizontal amplitude u_x and we are left with the three component column matrix of (4.17), where we relabel (y, z, φ) as $(1, 2, 3)$, respectively. We also assume that we have effected a 2D Fourier transform, which leaves us with a κ-dependent 3×3 differential system in z. To be precise, we concentrate on problems of the form

$$\mathbf{D}_\kappa(z) \cdot \mathbf{F}_\kappa(z) = 0 \; ; \; \mathbf{D}_\kappa(z) \cdot \mathbf{G}_\kappa(z, z') = \mathbf{I} \, \delta(z - z') \qquad (4.47)$$

with appropriate asymptotic and matching boundary conditions. The problem in general involves n independent components and then $\mathbf{I}$ in (4.47) is the $n \times n$ unit matrix.

The point is that, while the structure has mirror image symmetry, the n-component solution usually does not have a definite parity under the inversion $z \to -z$. For instance, in the case of the mixed polar optical modes each component, separately considered, does have a definite parity, but that of u_z is opposite to that of u_y and φ [2, 5].

Let us consider the first of equations (4.47). From the study of the transformation properties of the differential system we find that there is some operator $\hat{O}(z)$ which leaves $\mathbf{D}_K(z)$ invariant and under which $\mathbf{F}_K(z)$ has a definite parity p, i.e.

$$\hat{O}(z)\mathbf{D}_K(z)\hat{O}^{-1}(z) = \mathbf{D}_K(z),$$
$$\hat{O}(z)\mathbf{F}_K(z) = p\mathbf{F}_K(z),$$
$$(p = \pm 1). \qquad (4.48)$$

It then follows from general theorems in the theory of finite groups [8] that

$$\mathbf{G}_{K,p}(z, z') = \mathbf{G}_K(z, z') + p\hat{O}(z)\mathbf{G}_K(z, z') \qquad (4.49)$$

has definite parity $p = \pm 1$, as a function of z, under the same operator.

The precise form of this operator depends on the problem under study. For the moment we note that this analysis holds for the class of problems for which such operator exists and is not simply the inversion $z \to -z$. Then, from the two equalities contained in (4.49) we have

$$\mathbf{G}_K = \frac{1}{2}\left(\mathbf{G}_{K,+1} + \mathbf{G}_{K,-1}\right) \qquad (4.50)$$

which is a formal prescription for expressing $\mathbf{G}$ as the sum of its odd and even parts — in the general sense just explained. It follows that the total density of states is split as the sum of the two terms

$$\mathcal{N}_{K,p} = -\frac{1}{2\pi} Im \; Tr \; \mathbf{G}_{K,p}(\Omega; z, z) \qquad (4.51)$$

yielding separately the density of states with given parity. The practical usefulness of this will be seen in Chapter 5.

Now for the explicit form of the operator $\hat{O}$. In the physical problem here studied this is as follows:

$$\hat{O}(z)\mathbf{D}_K(z)\hat{O}^{-1} = \mathbf{M} \cdot \mathbf{D}_{-K}(-z) \cdot \mathbf{M}^{-1}, \qquad (4.52)$$

where $\mathbf{M}$ is the unitary matrix

$$\mathbf{M} = \begin{vmatrix} 1 & 0 & 0 \\ 0 & -1 & 0 \\ 0 & 0 & 1 \end{vmatrix}. \qquad (4.53)$$

In fact this is the form that it takes for many problems of physical interest [7] with a different $\mathbf{M}$ for each problem. We can define this as an inversion operator in a generalised sense: it involves not only changing $z \to -z$, but also inverting the sign of the *vector* κ, besides the intervention of the matrix **M**. *Parity* in this case means precisely that

$$\begin{aligned} \mathbf{M} \cdot \mathbf{D}_{-\kappa}(-z) \cdot \mathbf{M}^{-1} &= \mathbf{D}_{\kappa}(z), \\ \mathbf{M} \cdot \mathbf{F}_{-\kappa}(-z) &= p\mathbf{F}_{\kappa}(z) \qquad ; (p = \pm 1). \end{aligned} \qquad (4.54)$$

The need to invert also the sign of κ arises from the fact that there are second order crossed derivatives which in 2D Fourier transform introduce terms of the form $\kappa \partial/\partial z$ and the role of **M** is to change the sign of u_y, so all three components have then the same parity.

For the Surface Green Function Matching algebra it suffices to note that $\mathbf{G}_\kappa(z, z')$ transforms like $\mathbf{D}_\kappa$, so the transformed $\mathbf{G}_\kappa$ entering the r.h.s. of (4.49) is

$$\begin{aligned} \hat{O}(z)\mathbf{G}_\kappa(z, z')\hat{O}^{-1}(z) &= \mathbf{M} \cdot \mathbf{G}_{-\kappa}(-z, z') \cdot \mathbf{M}^{-1} \\ &= \mathbf{G}_\kappa(z, -z'). \end{aligned} \qquad (4.55)$$

The second equality is proved in Ref. [7] and is very useful for the practical implementation of the Surface Green Function Matching analysis.

The symmetric structure is thus separately studied in orthogonal subspaces for each parity, just as is usually done in terms of amplitudes. Spectral functions, eigenvalues and eigenfunctions — i.e. n-component amplitudes — are then obtained from separate Surface Green Function Matching calculations.

We thus have two optional alternatives to study the matching problem for one or more interfaces, so we can do it in terms of amplitudes, as in Chapter 3, or in terms of Green functions, as in the present chapter, according to practical convenience.

The physical nature of the polar optical modes in heterostructures

We now return to the physical model introduced in Chapters 2 and 3 and to the discussion of the restricted models initiated in section 3. In full 3D Fourier transform we introduce the longitudinal and transverse projectors

$$\mathbf{L} = \frac{1}{k^2}\mathbf{k}\mathbf{k} \quad ; \quad \mathbf{T} = \mathbf{I} - \mathbf{L}, \qquad (4.56)$$

so the model is then embodied in the field equation

$$\rho\left[\omega^2 - \omega_T^2 + \beta_L^2 k^2\right] \mathbf{L} \cdot \mathbf{u} + \alpha \mathbf{L} \cdot \mathbf{E} + \rho\left[\omega^2 - \omega_T^2 + \beta_T^2 k^2\right] \mathbf{T} \cdot \mathbf{u} = 0 \quad (4.57)$$

which, we recall, holds in the quasistatic limit $c \to \infty$.

In a bulk medium we can separate out longitudinal and transverse solutions, a definition which is established by reference to the mechanical amplitude, meaning that in an longitudinal/transverse solution $\nabla \times \mathbf{u}/\nabla \cdot \mathbf{u}$ — i.e. $\mathbf{T} \cdot \mathbf{u}/\mathbf{L} \cdot \mathbf{u}$ — vanishes. It is trivial to see that, since $\mathbf{E}$ is purely longitudinal, so are $\mathbf{D}$ and $\mathbf{P}$ and thus in the transverse modes there is only a mechanical vibration uncoupled to the electric field, which is only coupled to the longitudinal mechanical vibration. The asymptotic boundary conditions which apply to (4.56) for a bulk system do not in any way alter these facts.

The crucial feature common to several models — of the hydrodynamical or dielectric type — is *the assumption that this also holds for matched systems and that one can have purely longitudinal solutions for heterostructures*. This, however, is in general incorrect, as in problems involving vector fields the matching boundary conditions mix longitudinal and transverse polarisations ([1], section 9.5).

For isotropic models and with the geometry here chosen the u_x vibration factorises out as the purely mechanical transverse horizontal mode in any case, so we concentrate on the modes $\mathbf{F}$ of (4.17).

Now, *if* one assumes that $\mathbf{u}$ is purely longitudinal, with $\nabla \times \mathbf{u} = 0$, then one can write a simple relationship

$$\mathbf{E} = \frac{4\pi\alpha}{\epsilon_\infty} \mathbf{u} \quad (4.58)$$

and at this stage there are two options [9].

In the mechanical models of the hydrodynamical type one eliminates $\mathbf{E}$ and ends up with just one differential equation for one mechanical amplitude, namely

$$\left[\frac{d^2}{dz^2} + k_L^2\right] u_y = 0 \quad ; \quad k_L^2 = \frac{\omega_L^2 - \omega^2 - \beta_L^2 \kappa^2}{\beta_L^2}. \quad (4.59)$$

(The u_z component is then obtained from the vanishing of $\nabla \times \mathbf{u}$). In the so called *hydrodynamical model* [10] the problem actually is cast in a different way which is physically equivalent. For reasons soon to be seen we shall keep it in the form (4.59). The point is that the formulation of the problem then rests on a differential equation for a mechanical amplitude and one can only impose mechanical matching boundary conditions. One thus obtains $\mathbf{u}$ and when from it one calculates $\mathbf{E}$ then one finds that in general, for $\kappa \neq 0$, E_y is discontinuous and so is φ. Some ingenious *ad hoc* tricks have been proposed to get round this disturbing feature but the formal objection remains there. Furthermore, a stronger objection will be seen presently.

In the alternative choice one uses (4.58) to eliminate **u**. Then it is possible to write down a *longitudinal* dielectric function for the bulk media and the problem is formulated in terms of electrical boundary conditions. Electrostatic continuity is then achieved at the expense of mechanical *discontinuity*, for which one can also try some *ad hoc* artefact, but the formal objection remains again.

There is another important observation. The physical discussion focuses on the matching boundary conditions, but the model is embodied, to begin with, in the differential equation and here lies a significant difference between the two. Seen in full 3D Fourier transform, the inverse of the differential operator in the mechanical model is

$$\frac{1}{-k_z^2 + k_L^2}, \qquad (4.60)$$

while for the dielectric model we have

$$\frac{1}{(k_z^2 + \kappa^2)} \frac{\omega^2 - \omega_T^2 + \beta_L^2(k_z^2 + \kappa^2)}{\epsilon_\infty(k_z^2 - k_L^2)}. \qquad (4.61)$$

The z-dependence of the field amplitude is determined by the poles of the resolvent (4.60) and (4.61) in the complex k_z-plane. Both contain the poles at $\pm k_L$, which originate field amplitudes going like $exp(\pm i k_L z)$, but (4.61) contains another pole at $k_z = \pm i\kappa$. This originates terms going like $exp(-\kappa|z|)$ which can never be obtained from (4.60) or from any model based solely on mechanical field equations. Indeed this pole is of electrical origin and comes from the mathematical structure of the Poisson equation written in planar geometry. Thus the difference between the two types of models lies not only in the matching boundary conditions but also in the mathematical structure of the differential equation.

The situation is clearly unsatisfactory since starting from the same relationship (4.58) one arrives at a solution with either electrostatic or mechanical discontinuity. Moreover, the two models can never have amplitudes with a z-dependence of the same form, as has just be seen. The point is that (4.58) is only correct if one can really separate longitudinal and transverse solutions and this is in general incorrect for systems matching at interfaces. It is the assumption that a purely longitudinal solution exists that has confused the issue.

A phenomenological model which is intrinsically consistent and rid of these difficulties has been discussed in Chapter 3 in general terms. It is now interesting to see the physical nature of the solutions one obtains when this model is used to study heterostructures, for which the Surface Green Function Matching analysis provides a useful tool. In the case of a quantum well the attention focuses sometimes on the *confined* modes, which will be discussed in section 5. In a single interface, if the bulk phonon branches of the two constituent materials do not overlap and, as is often the case, there are only modes bounded to one side of the dividing surface. But all the heterostructures may also admit interface modes, with amplitudes localised in the neighbourhood of the surface and a weakness of the models of the hydrodynamical type is their inability to describe these — interface — modes.

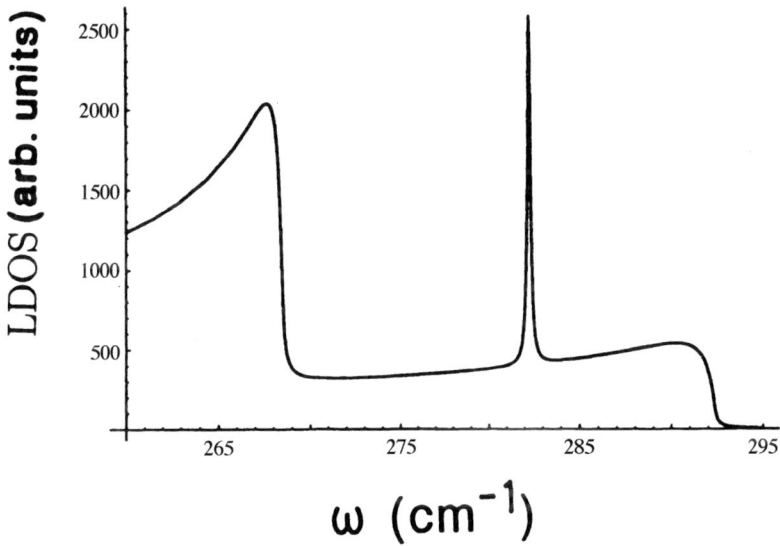

Figure 4.5 Local density of states in arbitrary units versus ω in cm^{-1}, for $\kappa = 0$, projected at the Al$_{0.9}$Ga$_{0.1}$As/GaAs interface. The frequency range spans the interval from just above $\omega_{L,GaAs}$ to a little below $\omega_{T,GaAs}$. A similar picture would describe the situation for the equivalent frequency range in Al$_{0.9}$Ga$_{0.1}$As.

The meaning of this was explained in a study of the Al$_{0.9}$Ga$_{0.1}$As system [9] by analysing in detail the spectral functions introduced in section 4. Figure 4.5 shows the local mode density at the location of the interface — $z = 0$ — for $\kappa = 0$ as a function of ω. The calculation focuses on the GaAs-type modes and so the figure spans the range of frequencies from somewhat below ω_T to somewhat above ω_L for GaAs. Proceeding from right to left we see a first sharp rise at ω_L and then another sharp rise at ω_T. Without the sharp peak in between this would be just the local projection of the GaAs bulk mode density reflecting the usual bulk threshold effects [1, 11], but there is this peak signalling the presence of an interface mode with a spectral strength which is high just when we project at the interface.

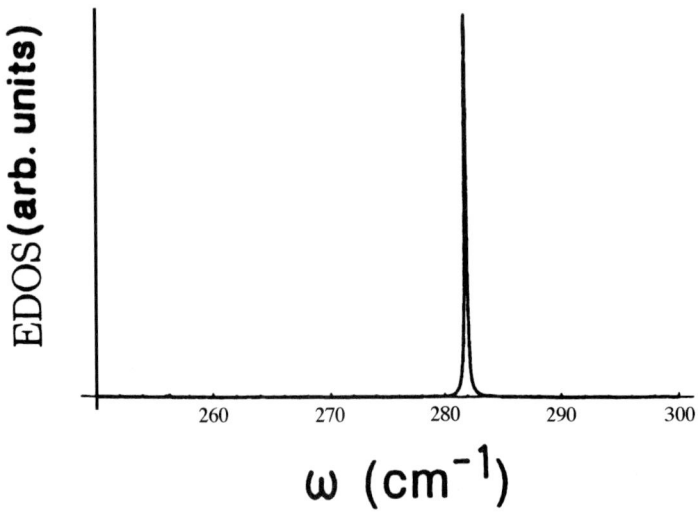

Figure 4.6 Out of the total spectral strengths adding up to the local density of states shown in Figure 4.5, the electrical spectral strength is separated out and shown here. All the electrical character lies in the interface mode, between ω_L and ω_T.

We can analyse the physical nature of this spectral strength by noting that, for the **F** modes of (4.17), in the sum over diagonal elements of the trace formula the first two terms contain the mechanical contribution while the electrical character is measured by the third term. Thus, by evaluating the contribution of these terms separately we determine the mechanical and/or electrical character contained in the total spectral strength of a given mode. Figure 4.6 shows the electrical part of the spectral strength for ω in the neighbourhood of the sharp peak of Figure 4.5. It is clear that the interface mode is in fact an electrical mode, which is physically obvious: The only coupling between two mechanically very different media is the electrostatic coupling.

Further insight into this can be gained by studying increasing values of κ. By following the position of the peak one then obtains the dispersion relation $\omega(\kappa)$ in a way which in fact is numerically more practical than looking for the roots of the secular determinant. Figure 4.7 was obtained for $\kappa = 2 \times 10^6$ cm^{-1}. Electrical (upper part) and mechanical (lower part) mode densities at the frequency of the eigenvalue $\omega(\kappa)$ for this value of κ were separately calculated as a function of z with GaAs on the right hand side. A small amplitude $|u_y| <<$

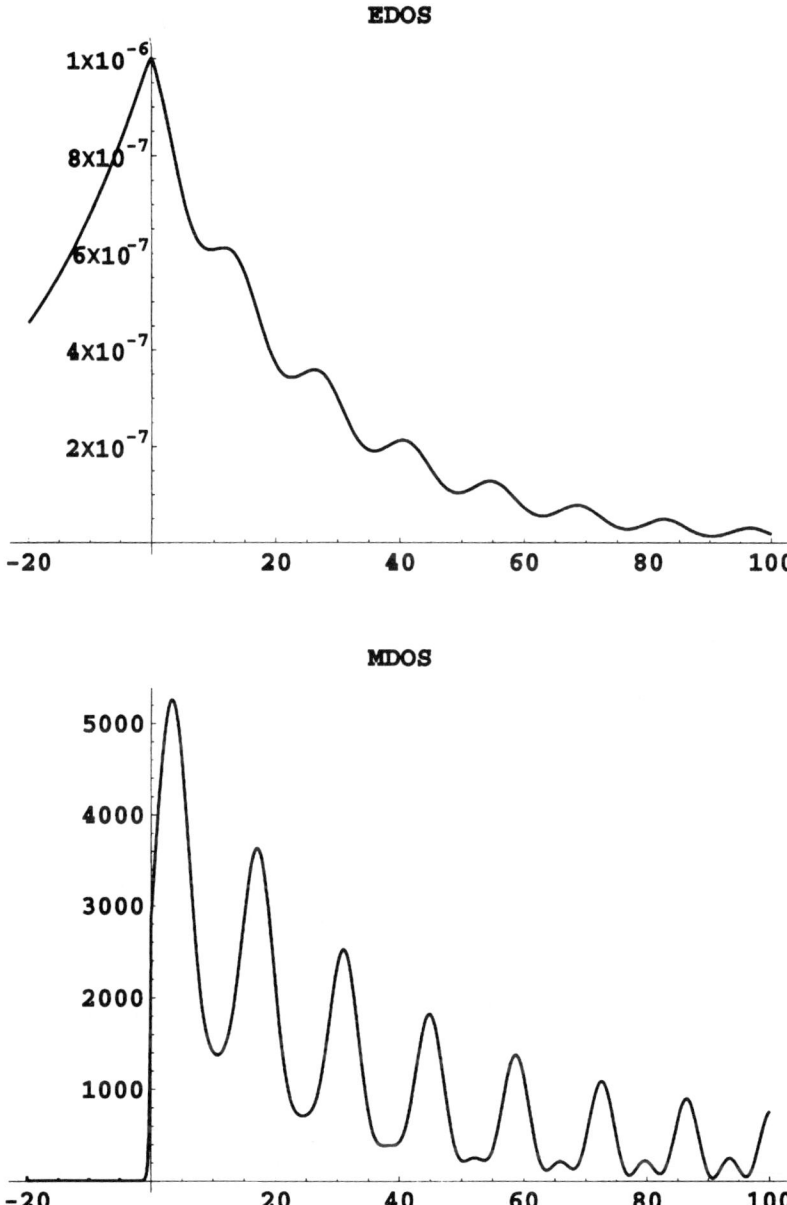

Figure 4.7 Spatial dependence of the electrical (upper) and mechanical (lower) spectral strength for the $Al_{0.9}Ga_{0.1}As/GaAs$ interface corresponding to $\kappa = 2 \times 10^6$ cm^{-1} exhibiting the amplitude accumulation near the interface which characterises an interface mode.

$|u_z|$ is not shown, so the figure shows $|\varphi(z)|^2$ and $|u_z(z)|^2$. The electrostatic potential decays like $exp(-\kappa|z|)$, corresponding to the pole of electrical origin described above. The evanescent oscillations on the GaAs side are due to the fact that this mode is actually a resonance with the bulk GaAs continuum, but a very sharp one — Figure 4.5. The mechanical amplitude decays extremely strongly on the other side, as $Al_{0.9}Ga_{0.1}As$ is practically a mechanically rigid barrier for the GaAs modes, so u_z is practically nil for $z = 0$. It also decays into the GaAs bulk but it tends to a steady oscillation because of the resonance with the continuum. Nevertheless, since it partakes in the character of an interface mode, it has an accumulation of amplitude near the interface, where all the electrical spectral strength is accumulated.

The electrical nature of the interface mode explains why the dielectric models can account for this mode and this is often held as an argument in favour of the dielectric models. However the formal objection remains in full strength, the assumption that **u** is purely longitudinal is not correct and the mechanical discontinuity hidden in the solution is a pathology of the model. Moreover, even if one starts from bulk media described by one longitudinal dielectric function relating purely longitudinal **E** and **D** fields, this does not hold for matched systems, where it can be seen that **D** then has a singular transverse part going like $\delta(z)$ ([1], section 9.5).

Solutions with **u** approximately longitudinal can provide a reasonable approximation in some particular cases, but in general a correct formulation of the problem requires a complete phenomenological model of the type discussed in Chapters 2 and 3 allowing for full mixing of longitudinal and transverse components and coupling of **u** and φ fields. More on the physical nature of polar optical modes in heterostructures will be said in section 5.

Special matching conditions for coupled fields

As stressed above polar optical modes involve two coupled fields, **u** and φ. This situation appears in other problems of physical interest, such as piezoelectric modes (also coupling of mechanical and electrical fields but different constitutive relations and with acoustic type of mechanical vibrations) or magnetoelastic waves (acoustic vibrational field coupled to magnetic field). In a matching analysis this may sometimes require some special matching boundary conditions for one of the fields while full matching is imposed on the other one. We shall consider some situations of physical interest which appear in practice and in which the two coupled fields are **u** and φ, although the formal analysis holds equally for any two coupled fields.

It proves convenient in the following to use a more general notation in which $\mathbf{F}_M$ denotes the mechanical part of the full field — i.e. **u** — and $\mathbf{F}_E$ the electrical part. In the quasistatic limit, with no retardation effects, we are interested only in the scalar potential φ but the formal analysis would equally hold for a more general case accounting for the full electromagnetic field. The quantities appearing in the analysis are equally partitioned and so is the full unit, which is decomposed in the 'mechanical unit' $\mathbf{I}_M$ and the 'electrical unit' $\mathbf{I}_E$. For instance, in the example of (4.15), if we factorise out the purely mechanical transverse horizontal mode, we have

$$\mathbf{I}_M = \begin{vmatrix} 1 & 0 & 0 \\ 0 & 1 & 0 \\ 0 & 0 & 0 \end{vmatrix} \; ; \; \mathbf{I}_E = \begin{vmatrix} 0 & 0 & 0 \\ 0 & 0 & 0 \\ 0 & 0 & 1 \end{vmatrix} \; ; \; \mathbf{I} = \mathbf{I}_M + \mathbf{I}_E \; . \tag{4.62}$$

For other situations the full unit matrix could be larger than 3×3 and $\mathbf{I}_M$ and $\mathbf{I}_E$ would have to be correspondingly defined in an obvious manner, but the formalism is the same in all cases. The point is that $\mathbf{I}_M$ and $\mathbf{I}_E$ have the nature of projectors onto their corresponding subspaces and their formal use is a very practical device to separate out the mechanical and electrical parts.

The basic equations are then, in Fourier transform, of the general form

$$\begin{vmatrix} \hat{\mathbf{D}}_{MM} & \hat{\mathbf{D}}_{ME} \\ \hat{\mathbf{D}}_{EM} & \hat{\mathbf{D}}_{EE} \end{vmatrix} \cdot \begin{vmatrix} \mathbf{F}_M \\ \mathbf{F}_E \end{vmatrix} = \mathbf{0} \; ,$$

$$\begin{vmatrix} \hat{\mathbf{D}}_{MM} & \hat{\mathbf{D}}_{ME} \\ \hat{\mathbf{D}}_{EM} & \hat{\mathbf{D}}_{EE} \end{vmatrix} \cdot \begin{vmatrix} \mathbf{G}_{MM} & \mathbf{G}_{ME} \\ \mathbf{G}_{EM} & \mathbf{G}_{EE} \end{vmatrix} = \mathbf{I} \; . \tag{4.63}$$

In the example of (4.15), having factorised D_{xx} and G_{xx} out, we have

$$\hat{\mathbf{D}}_{MM} = \begin{vmatrix} \hat{D}_{yy} & \hat{D}_{yz} \\ \hat{D}_{zy} & \hat{D}_{zz} \end{vmatrix} \; ; \; \hat{\mathbf{D}}_{ME} = \begin{vmatrix} \hat{D}_{y\varphi} \\ \hat{D}_{z\varphi} \end{vmatrix} ,$$

$$\hat{\mathbf{D}}_{EM} = \begin{vmatrix} \hat{D}_{\varphi y} & \hat{D}_{\varphi z} \end{vmatrix} \; ; \; \hat{\mathbf{D}}_{MM} = \hat{D}_{\varphi\varphi} \; . \tag{4.64}$$

The three cases we shall study are:

1. *Free surface.* This is selfexplanatory and corresponds to the semiinfinite medium, either for polar optical modes in polar semiconductors — or dielectrics — or else for piezoelectrics. For problems involving one field this is the standard free surface case; the distinctive feature here is that free surface conditions hold only for $\mathbf{F}_M$, while $\mathbf{F}_E$ is subject to standard full matching conditions.

2. *Hard mechanical wall*. This situation arises when the frequency ranges of the optical bulk bands of the two media meeting at the interface are wide apart. For instance, the difference between the highest (longitudinal) GaAs optical frequency and the lowest (transverse) AlAs optical frequency is larger than the with of the bulk bands. Then any GaAs mode reaching the GaAs-AlAs interface is in practice totally reflected so that, seen from the GaAs side, the interface can be treated as an infinitely rigid mechanical wall, while the electrostatic field penetrates everywhere. This is actually often the case in practice. For instance, in GaAs-based heterostructures this approximation holds good with the ternary compound $Al_xGa_{1-x}As$ for values of x from 1 to about 0.6.

3. *Electrostatically grounded surface*. This is a case of great practical interest in piezoelectrics [11]. The surface is coated with a very thin metallic film, so thin that it does not affect the free mechanical vibration but it provides a sufficient metallic contact that the surface can be grounded to a fixed potential. This is often done, as it turns out to have practical advantages for surface wave piezoelectric devices [11].

In all these cases the two coupled fields satisfy different types of boundary conditions at the matching surface and the formalism outlined in section 4 requires appropriate modification. We shall see the form that the Surface Green Function Matching analysis takes in such cases.

Case 1: *Free surface*

By convention the vacuum is on side 1. Then $\mathbf{G}_1$ has only the electrical part, but in order to carry out a matching of the two unequal media we must formally put all the quantities together in the same matrix format spanning the M and E subspaces. Thus we define the full $\mathbf{G}_1$ matrix as having three identically nil submatrices. These are the MM, ME and EM submatrices, so only the EE submatrix is nonvanishing and the form of $\mathbf{G}_1$, as well as that of $\mathcal{G}_1$ — surface projection — is

$$\begin{vmatrix} 0 & 0 \\ 0 & * \end{vmatrix}, \qquad (4.65)$$

where the asterisk denotes that this element simply does not vanish.

A formal way of putting this into the analysis is to express $\mathbf{G}_1$, $\mathcal{G}_1$, etc, as, say, $\mathbf{I}_E \mathbf{G}_1 \mathbf{I}_E$. Then all the algebraic operations pertaining to $\mathbf{G}_1$, such as the inversion of $\mathcal{G}_1$, are carried out in the E subspace and then the result is cast in the large matrix format (4.65). Thus writing down $\mathcal{G}_1^{-1}$ does *not* imply inverting a singular matrix.

Now, an elementary excitation incident on the surface form side 2 has both, the M and E parts. Of these, only the E part propagates outside and the transfer matrix which propagates this amplitudes is

$$\mathbf{I}_E \cdot \mathbf{G}_1(z, 0) \cdot \mathbf{I}_E \cdot \mathcal{G}_1^{-1} \cdot \mathbf{I}_E . \qquad (4.66)$$

This is the canonical term to be employed in the general Surface Green Function Matching formulae of section 4 concerning side 1, while the quantities pertaining to side 2 involve $\mathbf{G}_2$, $\mathcal{G}_2$ and $\mathcal{G}_2^{-1}$ in the standard way, all four submatrices then being nonvanishing. One can now differentiate (4.66) with respect to z and carry out the Surface Green Function Matching programme in the usual way. In this process one encounters the standard linear differential form $\mathcal{A}^\pm$ which for side 1 takes again the form (4.65).

Then defining

$$\mathcal{B}^\pm \equiv \mathbf{I}_E \cdot \mathcal{A}^\pm \cdot \mathbf{I}_E , \qquad (4.67)$$

the matching formula for the full $\mathcal{G}_s^{-1}$ is

$$\mathcal{G}_s^{-1} = \mathbf{I}_E \cdot \mathcal{B}_1^{(+)} \cdot \mathbf{I}_E \cdot \mathcal{G}_1^{-1} \cdot \mathbf{I}_E - \mathcal{A}_2^{(-)} \cdot \mathcal{G}_2^{-1} . \qquad (4.68)$$

The formal prescription for the evaluation of a term like $\mathbf{G}_1(z, 0) \cdot \mathcal{G}_1^{-1} \cdot \mathcal{G}_s$ is then

$$\mathbf{I}_E \cdot \mathbf{G}_1(z, 0) \mathbf{I}_E \cdot \mathcal{G}_1^{-1} \cdot \mathbf{I}_E \cdot \left[\mathbf{I}_E \cdot \mathcal{B}_1^{(+)} \cdot \mathbf{I}_E \mathcal{G}_1^{-1} \cdot \mathbf{I}_E - \mathcal{A}_2^{(-)} \cdot \mathcal{G}_2^{-1} \right] , \qquad (4.69)$$

which precisely prescribes all the terms and contributions surviving in the end. The full Surface Green Function Matching program can then be carried out with no ambiguity.

Case 2: Hard mechanical wall

We consider an interface like GaAs-AlAs. Irrespective of numerics the standard Surface Green Function Matching analysis can be carried out and leads to the fully matched $\mathbf{G}_s$. Now, as explained above, looking at the problem from the GaAs side the interface can be treated as a hard wall which is infinitely rigid for the mechanical vibration. Let $\mathbf{G}_h$ denote the matched GF obtained by applying hard wall matching boundary conditions for the mechanical amplitude, i.e.

$$\mathbf{F}_M(z) = 0 \; ; \; (0 \leq z), \qquad (4.70)$$

while full matching is imposed on $\mathbf{F}_E$. The surviving part of the matching boundary conditions expresses only electrostatic matching but it does so in exactly the same manner as in the general case. Indeed it is easy to prove formally that

$$\mathbf{I}_E \cdot \mathcal{G}_h^{-1} = \mathbf{I}_E \cdot \mathcal{G}_s^{-1}, \tag{4.71}$$

while the vanishing of the mechanical amplitude at the matching surface requires that

$$\mathbf{I}_M \cdot \mathcal{G}_s = 0. \tag{4.72}$$

The inverse relationship between $\mathcal{G}_h^{-1}$ and $\mathcal{G}_h$ is defined by the condition

$$\mathcal{G}_h \cdot \mathcal{G}_h^{-1} = \mathbf{I}_E. \tag{4.73}$$

It follows from (4.71) through (4.73) that if we put

$$\mathcal{G}_s^{-1} = \Gamma = \begin{vmatrix} \Gamma_{MM} & \Gamma_{ME} \\ \Gamma_{EM} & \Gamma_{EE} \end{vmatrix} \tag{4.74}$$

then

$$\mathcal{G}_h^{-1} = \begin{vmatrix} 0 & 0 \\ 0 & \Gamma_{EE} \end{vmatrix}; \quad \mathcal{G}_h = \begin{vmatrix} 0 & 0 \\ 0 & \Gamma_{EE}^{-1} \end{vmatrix}. \tag{4.75}$$

Note that the secular determinant in this case is

$$det \left| \mathbf{I}_E \cdot \mathcal{G}_s^{-1} \cdot \mathbf{I}_E \right| = 0. \tag{4.76}$$

In all these equations $\mathcal{G}_s$ is given by the general matching formula (4.12) and then $\mathbf{I}_E$ and $\mathbf{I}_M$ pick out the E and M parts as they must intervene in the new matching analysis. Note that the results differ from those obtained for the free surface, although in both cases $\mathbf{F}_M$ does not propagates outside the medium we started from. We have seen how this is incorporated in the formalism for the vacuum. For the rigid medium the physical situation is the same, but the formal prescription is different: The medium has a full **G** matrix as in (4.63), but for frequency ranges away from its allowed bulk bands no normal modes propagates; the transfer matrix (4.66) yields an evanescent amplitude also for the $\mathbf{F}_E$ part.

Case 3: Electrostatically grounded surface

The case of the grounded surface — or interface — can also be studied in a similar manner in terms of the $\mathbf{I}_M$ and $\mathbf{I}_E$ projectors. Details can be found in Refs. [1, 12].

Let $\mathbf{G}_s$ be the full Green function for either the free surface or the interface and $\mathbf{G}_g$ that corresponding to the grounded surface (or interface). Then the mechanical matching is done in the same way for $\mathcal{G}_s^{-1}$ as for $\mathcal{G}_g^{-1}$ and one can prove that

$$\mathbf{I}_M \cdot \mathcal{G}_g^{-1} = \mathbf{I}_M \cdot \mathcal{G}_s^{-1} \,. \tag{4.77}$$

The situation is in a way the counterpart of that expressed by (4.71), but there is one formal difference in that here $\mathbf{I}_E \cdot \mathcal{G}_s$ need not vanish, as the surface potential need not be zero but, whatever the reference level chosen, it must be equal to some arbitrary constant without dynamical structure. It follows that $\mathcal{G}_g^{-1}$ is of the form

$$\mathcal{G}_g^{-1} = \begin{vmatrix} \Gamma_{MM} & \Gamma_{ME} \\ 0 & \mathbf{X} \end{vmatrix}, \tag{4.78}$$

where $\mathbf{X}$ is some arbitrary constant. Γ_{ME} is also proportional to $\mathbf{X}$, but its details are irrelevant. Note that the secular equation is now

$$det \left| \mathbf{I}_M \cdot \mathcal{G}_s^{-1} \cdot \mathbf{I}_M \right|, \tag{4.79}$$

in correspondence with (4.76) for the hard wall case. Furthermore, the phase function is

$$arg\, det \left| \mathcal{G}_g^{-1} \right| = arg\, det \left| \Gamma_{MM} \right| + arg\, \mathbf{X} \,. \tag{4.80}$$

Since $\mathbf{X}$ is a constant with no dynamical structure its argument vanishes (or is independent of ω). Thus, since the mode density is in general obtained from the derivative of the phase function also for the matching problem, just as in scattering theory [1, 12–14], the arbitrary constant is seen to be physically irrelevant, as it must be, since no physical result depends on it.

The above considerations indicate how to proceed when a matching problem involves two coupled fields and the matching boundary conditions imposed on these are different.

Having seen how to deal with one single surface (or interface) it is obvious how to proceed when more interfaces are involved. For instance, for a GaAs quantum well with AlAs barriers we have two interfaces and mechanically rigid walls at both. The analysis can then be carried out by combining the arguments given here with those of section 4.

References

1. F. García Moliner and V.R. Velasco, *Theory of Single and Multiple Interfaces: The Method of Surface Green Function Matching*. World Scientific, Singapore (1992).

2. R. Pérez-Alvarez, F. García-Moliner, V.R. Velasco, C. Trallero-Giner, Phys. Rev. **B48**, 5672 (1993). A. Chubykalo, V.R. Velasco and F. García-Moliner, Surf. Sci. **319**, 184 (1994).

3. E. Dieulessaint and D. Royer, *Elastic Waves in Solids*, J. Wiley, New York (1980).

4. N. Zou, K.A. Chao, Phys. Rev. Lett. **69**, 3224 (1992).

5. R. Pérez-Alvarez, F. García-Moliner and V.R. Velasco, J. Phys.: Condensed Matter **7**, 2037 (1995).

6. S. Vlaev, V.R. Velasco and F. García-Moliner, Phys. Rev. **B49**, 1122 (1994).

7. R. Pérez-Alvarez, F. García-Moliner, H. Rodríguez-Coppola and V.R. Velasco, Surf. Sci. **371**, 455 (1997).

8. M. Hammermesh *Group theory and its application to physical problems*, Addison-Wesley, London (1964).

9. F. García-Moliner, in *Phonons in nanostructures*, J.-P. Leburton, C. Sotomayor and J. Pascual edits., Kluwer Academic Publishers, Netherlands (1993).

10. M. Babiker, J. Phys. C:Condensed Matter **19**, 683 (1986).

11. E. Dieulessaint and D. Roger, in *Handbook of surfaces and interfaces* Vol. 2, p. 65, L. Dobrzynsky, edit. Garland, New York, (1983).

12. V.R. Velasco and F. García-Moliner, Surf. Sci. **143**, 93 (1984).

13. L. Dobrzynsky, V.R. Velasco and F. García-Moliner, Phys. Rev. **B35**, 5872 (1987).

14. F. García-Moliner and F. Flores, *Introduction to the theory of solid surfaces*. Cambridge University Press (1979).

CHAPTER FIVE

Polar optical modes in layered structures

Having at our disposal the general techniques for matching polar optical modes either in terms of linearly independent solutions or in terms of Green functions, we can now use these alternatively according to convenience to study various heterostructures of interest, namely, quantum wells — or barriers — superlattices and the type of double barrier structures used in resonant tunnelling experiments, where phonon replicas evidence inelastic phonon scattering of the tunnelling electrons.

Finally, most of the heterostructures studied — in this book as well as elsewhere — are made by combining materials having a common ion, for instance As in GaAs/AlAs or GaAs/Al$_x$Ga$_{1-x}$As heterostructures. However, interfaces with no common ion are receiving increasing attention and we also present a brief discussion of this problem, with an example of application.

Linearly Independent Solutions

For a given homogeneous bulk medium (3.6) and (3.7) can be solved with the help of the auxiliary potentials Ψ and Γ introduced in section 3. Thus we seek for **u** a solution of the form:

$$\mathbf{u} = \nabla\Psi + \nabla \times \mathbf{\Gamma} , \qquad (5.1)$$

where the supplementary condition $\nabla \cdot \mathbf{\Gamma} = 0$ is imposed. It can be easily verified, after substitution of (5.1) in (3.6) and (3.7), that Ψ and Γ satisfy the equations:

$$\nabla^2 \left[\nabla^2 \mathbf{\Gamma} + \frac{\omega_T^2 - \omega^2}{\beta_T^2} \mathbf{\Gamma} \right] = 0 , \qquad (5.2)$$

$$\nabla^2 \left[\nabla^2 \Psi + \frac{\omega_L^2 - \omega^2}{\beta_L^2} \Psi \right] = 0 , \qquad (5.3)$$

which can be solved by elementary methods. Then, having solved (5.2) and (5.3) we obtain $\mathbf{u}(\mathbf{r})$ and, after 2D Fourier transform in the (x, y) plane we have a solution of the form

$$\mathbf{u}(\mathbf{r}) = \mathbf{u}(z)\, e^{i\boldsymbol{\kappa}\cdot\boldsymbol{\rho}} .\qquad(5.4)$$

The function $\varphi(\mathbf{r})$ is also of the form

$$\varphi(\mathbf{r}) = \varphi(z)\, e^{i\boldsymbol{\kappa}\cdot\boldsymbol{\rho}} \qquad(5.5)$$

and the problem is to find $\varphi(z)$ which, using (5.4) and (5.5) in (3.7), satisfies the ordinary differential equation

$$\left(\frac{d^2}{dz^2} - \kappa^2\right)\varphi(z) = \frac{4\pi\alpha}{\epsilon_\infty}\left[i\kappa_x u_x + i\kappa_y u_y + \frac{du_z}{dz}\right],\qquad(5.6)$$

where u_x, u_y and u_z are already known. The general solution of (5.6) can thus be found.

An important symmetry of the present problem is related to the invariance of the system under arbitrary rotations about the z axis as a consequence of the assumed isotropy of the constituent media. Therefore the x, y axes can be chosen in any convenient way without loss of generality. We take the y axis along the vector $\boldsymbol{\kappa}$, i.e. $\boldsymbol{\kappa} = (0, \kappa_y) = (0, \kappa)$. Then two fundamental types of linearly independent solutions are obtained.

One is a completely decoupled purely transverse solution of the form given by the one-column matrices:

$$\left| \begin{array}{c} i\frac{\kappa^2+k_T^2}{k_T}\cos(k_T z) \\ 0 \\ 0 \\ 0 \end{array} \right| \;;\; \left| \begin{array}{c} -\frac{\kappa^2+k_T^2}{k_T}\sin(k_T z) \\ 0 \\ 0 \\ 0 \end{array} \right| .\qquad(5.7)$$

It is clear that solutions (5.7) have definite parity as a consequence of the system invariance under the transformation of $z \to -z$.

The other linearly independent solutions involve the three amplitudes u_y, u_z and φ, all coupled. Each one of these must have independently a definite parity under a change of sign of z and it is seen from (5.6) that u_y and φ have the same parity, which is opposite to that of u_z. We are encountering again the same situation we discussed in the Green function study of symmetric structures in Chapter 4. The basis of the six linearly independent solutions for the subspace of u_y, u_z and φ can be arranged as consisting of two subsets, one formed by the three column matrices

$$\begin{vmatrix} 0 \\ -\kappa \sin(k_L z) \\ ik_L \cos(k_L z) \\ i\frac{4\pi\alpha}{\epsilon_\infty} \sin(k_L z) \end{vmatrix} \;;\; \begin{vmatrix} 0 \\ k_T \sin(k_T z) \\ i\kappa \cos(k_T z) \\ 0 \end{vmatrix} \;;\; \begin{vmatrix} 0 \\ i\frac{\alpha\kappa}{\rho(\omega^2-\omega_T^2)} \sinh(\kappa z) \\ \frac{\alpha\kappa}{\rho(\omega^2-\omega_T^2)} \cosh(\kappa z) \\ \sinh(\kappa z) \end{vmatrix} , \quad (5.8)$$

where φ is odd, and the other formed by

$$\begin{vmatrix} 0 \\ i\kappa \cos(k_L z) \\ -k_L \sin(k_L z) \\ \frac{4\pi\alpha}{\epsilon_\infty} \cos(k_L z) \end{vmatrix} \;;\; \begin{vmatrix} 0 \\ -ik_T \cos(k_T z) \\ -\kappa \sin(k_T z) \\ 0 \end{vmatrix} \;;\; \begin{vmatrix} 0 \\ i\frac{\alpha\kappa}{\rho(\omega^2-\omega_T^2)} \cosh(\kappa z) \\ \frac{\alpha\kappa}{\rho(\omega^2-\omega_T^2)} \sinh(\kappa z) \\ \cosh(\kappa z) \end{vmatrix} , (5.9)$$

where φ is even.
Here we recall that

$$k_L = \sqrt{\frac{\omega_L^2 - \omega^2}{\beta_L^2} - \kappa^2} \;;\; k_T = \sqrt{\frac{\omega_T^2 - \omega^2}{\beta_T^2} - \kappa^2} . \qquad (5.10)$$

It is easy to recognise in terms of the amplitudes (5.8) and (5.9) the symmetry under the generalised inversion operator of Chapter 4.

It should be noted that there is a certain interval of ω ($\omega_T < \omega < \omega_L$) where k_T becomes pure imaginary. We are restricting ourselves to the long wavelength limit and hence the values of κ span just a fraction of the Brillouin Zone ($\sim 10\%$). Thus k_L is essentially real while k_T could be imaginary or real. We have thus found (through (5.7), (5.8) and (5.9)) a basis for the solution space with the advantage that each group leads to a subspace not mixed with the other one.

Quantum wells

Equation of motion, boundary conditions and matching conditions

In this section we focus our attention on the optical modes of GaAs-Al$_x$Ga$_{1-x}$As ($0 \leq x \leq 1$) quantum wells. In studying these systems we must assign to the ternary alloy the values of the frequencies for the longitudinal optical and transverse optical modes given in section 1 for the GaAs (AlAs) like modes in this alloy [1]. The β_L and β_T parameters were estimated from the experimental curves of Wang et al. [2] (see section 1).

Following the general arguments of Chapter 3 we maintain the usual asymptotic boundary conditions of regularity at $\pm\infty$ while imposing complete matching boundary conditions at the interfaces. We recall that, as discussed in Chapters 3 and 4, in some cases, such as the GaAs-AlAs interfaces, a good approximation is to assume a mechanically rigid wall, which simplifies the calculations. Indeed, as will be presently seen, if we solve the problem with full matching conditions then we find numerically that the penetration of the mechanical amplitude into the 'other' side is often in practice very quickly damped.

Now consider a quantum well type of structure of width d, concentrate on the modes of the material inside and assume that the rigid wall approximation holds. Applying this to a linear combination of the linearly independent solutions described in (5.8) we obtain the odd potential modes with $\varphi(z)$ odd, for the coupled oscillation modes. For the sake of simplicity we have assumed a uniform dielectric constant ϵ_∞ throughout the whole system. Numerical checks indicate that the differences in the crystal background dielectric constant ϵ_∞ are in practice unimportant. Then

$$\begin{aligned} u_y(z) &= A\,[C_1\,\sin(k_L z) + C_2\,\sin(k_T z) + C_3\,\sinh(\kappa z)]\ ,\\ u_z(z) &= A\,[C_4\,\cos(k_L z) + C_5\,\cos(k_T z) + C_6\,\cosh(\kappa z)]\ ,\\ \varphi(z) &= A\,[C_7\,\sin(k_L z) + C_8\,\sin(k_T z) + C_9\,\sinh(\kappa z)]\ , \quad (5.11) \end{aligned}$$

for $|z| < d/2$, while for $|z| > d/2$ we have

$$u_y(z) = 0\ ;\quad u_z(z) = 0\ ;\quad \varphi(z) = A\,C_{10}\,e^{-\kappa|z|}\,z/|z|\ , \qquad (5.12)$$

where A is the normalisation constant and

$$\begin{aligned} C_1 &= -\eta_\kappa\,i\,e_5\ ;\quad C_2 = \eta_T\,i\,e_3\ ;\quad C_3 = \eta_\kappa\,i\,e^{-\eta_\kappa}\,e_1\,e_2\,e_5\ ,\\ C_4 &= -\eta_L\,e_5\ ;\quad C_5 = -\eta_\kappa\,e_3\ ;\quad C_6 = \eta_\kappa\,e^{-\eta_\kappa}\,e_1\,e_2\,e_5\ ,\\ C_7 &= -\frac{2\pi\alpha d}{\epsilon_\infty}\,e_5\ ;\quad C_8 = 0\quad ;\quad C_9 = \frac{2\pi\alpha d}{\epsilon_\infty}\,e^{-\eta_\kappa}\,e_2\,e_5\ ,\\ & \hspace{6.5em} C_{10} = \frac{2\pi\alpha d}{\epsilon_\infty}\,e_4\,e_5\ , \quad (5.13) \end{aligned}$$

$$e_1 = \frac{\omega_L^2 - \omega_T^2}{\omega^2 - \omega_T^2}\ ,$$

$$e_2 = \sin\eta_L + \frac{\eta_L}{\eta_\kappa}\cos\eta_L,$$

$$e_3 = \eta_\kappa \sin\eta_L \cosh\eta_\kappa - \eta_L \cos\eta_L \sinh\eta_\kappa,$$

$$e_4 = \frac{\eta_L}{\eta_\kappa}\cos\eta_L \sinh\eta_\kappa - \sin\eta_L \cosh\eta_\kappa,$$

$$e_5 = \eta_\kappa \cos\eta_T \sinh\eta_\kappa + \eta_T \sin\eta_T \cosh\eta_\kappa, \quad (5.14)$$

$$\eta_L = k_L d/2 \quad ; \quad \eta_T = k_T d/2 \quad ; \quad \eta_\kappa = \kappa d/2. \quad (5.15)$$

For these states the following secular equation is found:

$$(\omega_L^2 - \omega^2)(\eta_L \cos\eta_L + \eta_\kappa \sin\eta_L)(\eta_T \sin\eta_T \cosh\eta_\kappa + \eta_\kappa \cos\eta_T \sinh\eta_\kappa) =$$
$$(\omega_T^2 - \omega^2)(\eta_\kappa \cos\eta_T - \eta_T \sin\eta_T)(\eta_L \cos\eta_L \sinh\eta_\kappa - \eta_\kappa \sin\eta_L \cosh\eta_\kappa).$$
$$(5.16)$$

This yields the dispersion relation $\omega(\kappa)$. From the linearly independent solutions described in (5.9) we obtain the even potential modes, with $\varphi(z)$ even:

$$u_y(z) = A'\left[C_1' \cos(k_L z) + C_2' \cos(k_T z) + C_3' \cosh(\kappa z)\right],$$
$$u_z(z) = A'\left[C_4' \sin(k_L z) + C_5' \sin(k_T z) + C_6' \sinh(\kappa z)\right],$$
$$\varphi(z) = A'\left[C_7' \cos(k_L z) + C_8' \cos(k_T z) + C_9' \cosh(\kappa z)\right], \quad (5.17)$$

for $|z| < d/2$, while

$$u_y(z) = 0 \quad ; \quad u_z(z) = 0 \quad ; \quad \varphi(z) = A' C_{10}' \, e^{-\kappa|z|}, \quad (5.18)$$

for $|z| > d/2$. Furthermore

$$C_1' = -\eta_\kappa e_5' \quad ; \quad C_2' = \eta_T e_3' \quad ; \quad C_3' = -\eta_\kappa \, e^{-\eta_\kappa} e_1' e_2' e_5',$$
$$C_4' = -i\eta_L e_5' \quad ; \quad C_5' = -i\eta_\kappa e_3' \quad ; \quad C_6' = i\eta_\kappa \, e^{-\eta_\kappa} e_1' e_2' e_5',$$
$$C_7' = \frac{2i\pi\alpha d}{\epsilon_\infty} e_5' \; ; \quad C_8' = 0 \quad ; \quad C_9' = \frac{2i\pi\alpha d}{\epsilon_\infty} \, e^{-\eta_\kappa} e_2' e_5',$$
$$C_{10}' = \frac{2i\pi\alpha d}{\epsilon_\infty} e_4' e_5' \quad , \quad (5.19)$$

$$e_1' = \frac{\omega_L^2 - \omega_T^2}{\omega^2 - \omega_T^2},$$

$$e'_2 = \frac{\eta_L}{\eta_\kappa} \sin \eta_L - \cos \eta_L ,$$

$$e'_3 = \eta_\kappa \cos \eta_L \sinh \eta_\kappa + \eta_L \sin \eta_L \cosh \eta_\kappa ,$$

$$e'_4 = \frac{\eta_L}{\eta_\kappa} \sin \eta_L \cosh \eta_\kappa + \cos \eta_L \sinh \eta_\kappa ,$$

$$e'_5 = \eta_T \cos \eta_T \sinh \eta_\kappa - \eta_\kappa \sin \eta_T \cosh \eta_\kappa . \qquad (5.20)$$

A' is a normalisation constant.

The secular equation is then

$$(\omega_L^2 - \omega^2)(\eta_\kappa \cos \eta_L - \eta_L \sin \eta_L)(\eta_T \cos \eta_T \sinh \eta_\kappa - \eta_\kappa \sin \eta_T \cosh \eta_\kappa) =$$
$$(\omega_T^2 - \omega^2)(\eta_\kappa \sin \eta_T + \eta_T \cos \eta_T)(\eta_L \sin \eta_L \cosh \eta_\kappa + \eta_\kappa \cos \eta_L \sinh \eta_\kappa)$$

$$(5.21)$$

It is important to realise that for both cases we have $\nabla \cdot \mathbf{u} \neq 0$ and $\nabla \times \mathbf{u} \neq \mathbf{0}$, and hence the coupled modes are neither purely longitudinal nor purely transverse, but mixed. We can only say that the modes may be predominantly longitudinal (quasi-longitudinal) in some cases or predominantly transverse (quasi-transverse) in other cases. In the next paragraph this point will be discussed with further details in a specific case. The influence of the electric potential φ leads to a certain character of 'interface' modes as can be seen by noting that $C_3, C'_3, C_6, C'_6 \neq 0$, but it is in order to remark that these are not purely interface modes. It is also interesting that $C_8 = C'_8 = 0$ and that u_y has always a $\pi/2$ phase difference with respect to u_z and φ.

For $\kappa = 0$ it can be easily verified that the coupled modes become uncoupled yielding a set of purely longitudinal modes with eigenfrequencies:

$$\omega_m^2 = \omega_L^2 - \beta_L^2 \left(\frac{2m\pi}{d}\right)^2 , \qquad (5.22)$$

and a set of purely transverse modes with eigenfrequencies:

$$\omega_m^2 = \omega_T^2 - \beta_T^2 \left(\frac{2m\pi}{d}\right)^2 . \qquad (5.23)$$

GaAs-AlAs double heterostructure

We now apply this analysis to a GaAs-AlAs double heterostructure. When GaAs/AlAs is inside, then the double heterostructure is a quantum well/barrier for the electronic problem. We concentrate on the modes inside and discuss dispersion relations and spatial dependence of vibration amplitudes and electrostatic potential as calculated with the rigid matching boundary conditions. We used the numerical values given in Table 1.1.

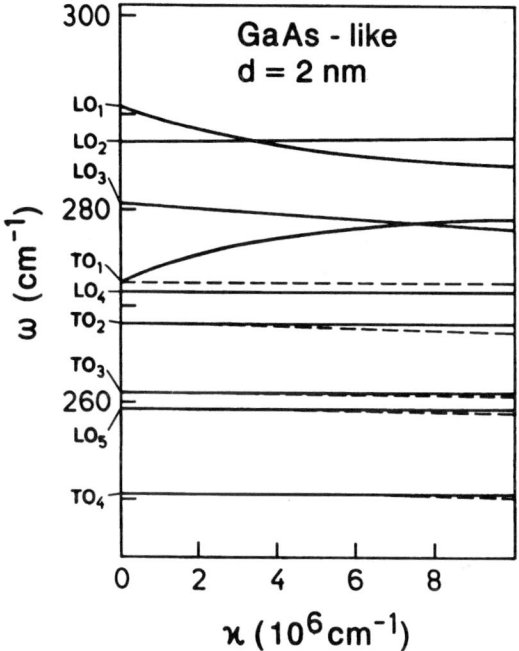

Figure 5.1 Dispersion branches — ω vs κ — for the GaAs-like normal modes in a 2 nm thick GaAs quantum well with AlAs barriers outside. The dashed lines indicate the shear horizontal -purely mechanical- modes, decoupled from the rest.

Figure 5.1 shows the dispersion relation of a GaAs quantum well with $d = 2$ nm. The labelling of the modes is indicated according to their character at $\kappa = 0$. It is a remarkable feature that the transverse and shear horizontal modes are practically degenerate at all values of κ except for the TO1 modes which have interface character. We keep the notation LOn and TOn to facilitate identification with the customary labelling, but we recall that the modes are purely longitudinal or transverse *only* for $\kappa = 0$. In the following discussion we shall sometimes refer to these branches as L or T for the sake of conciseness, but always keeping this reservation in mind. It is also clearly seen that branches with different parities cross each other while branches with the same parities originate gaps whenever they become close to each other. All branches are almost flat except in the intervals of κ where crossings and gaps take place. For higher frequencies the even potential modes display dispersion curves with a weak dependence on κ. It should be noted that for these frequencies the constants C_3, C_6, C_9 are large for odd potential modes. In the region of $\omega \sim \omega_T$ the opposite is true: then it is the dispersion curves corresponding to odd potential modes that are flat.

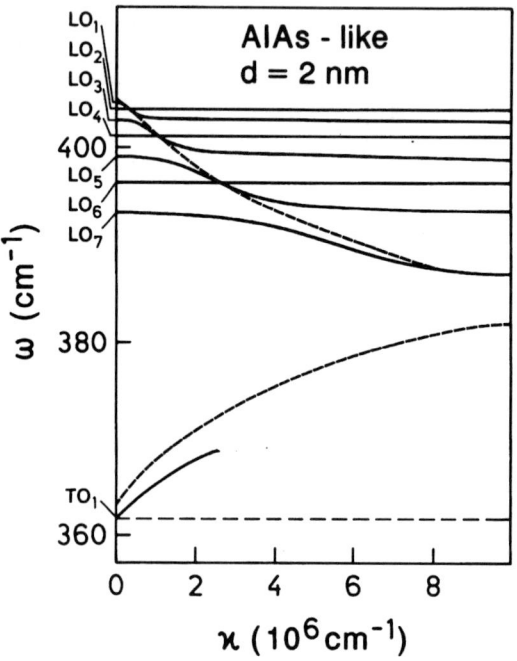

Figure 5.2 Dispersion branches — ω vs κ — for the AlAs-like normal modes in a 2 nm thick AlAs barrier with GaAs outside. The dashed lines indicate the shear horizontal — purely mechanical — modes, decoupled from the rest.

The intervals of κ showing more intense dispersion are those found to have a stronger electrical character according to the criteria discussed in section 4. Hence, the interface character is more prominent for such states. This stresses the point that there are no purely interface modes in this model, but for some intervals of κ the different modes can display predominantly interface character. Rigorously speaking, the presence of purely interface modes in other phenomenological approaches to the subject is a result of the unjustified decoupling of the longitudinal and transverse vibrations. For instance, in Refs. [3, 4] it is *a priori* assumed that the oscillations are pure longitudinal. When a dielectric continuum approximation is applied, as in [5, 6], purely interface modes are also obtained. It is important to emphasise that the correct treatment of the oscillation problem requires to solve coupled differential equations (such as Eqs. (3.6) and (3.7)). Hence, we can not achieve *purely* longitudinal, transverse or interface modes, but only oscillations with a mixed nature. However, the shortcomings of particularly restricted models have been discussed in Chapters 3 and 4, and we need not delve into this any further.

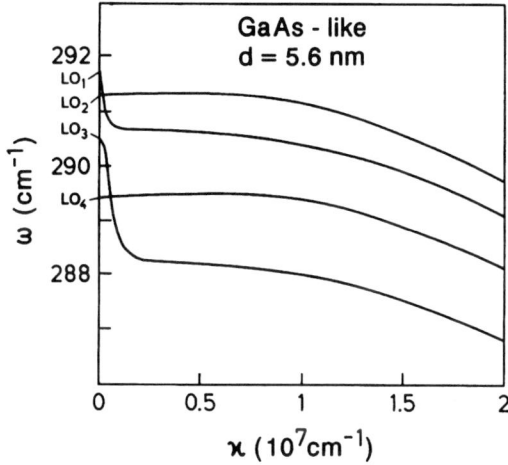

Figure 5.3 Dispersion branches — ω vs κ — for the first four GaAs-like modes in a 5.65 nm thick GaAs/AlAs quantum well.

The same features of the dispersion branches are present in Figure 5.2 for the case of an AlAs barrier with $d = 2$ nm. In this case we find a larger amount of modes in a similar interval of frequency; this is due to the particularly small value of the parameter β_L (the longitudinal branches are almost flat in the bulk material). Then the crossings and gaps (anticrossings) are more visible. As has just been stressed the ranges of the dispersion branches exhibiting a stronger dispersion are those having a predominant electrical character. It is interesting to note that if one links them up in a sort of envelope then the resulting curve resembles very closely the interface mode appearing in restricted models, which is in line with the physical discussion of section 4.

Figure 5.3 shows the dispersion relation of the first four modes of a 56 Å GaAs quantum well, with the usual crossings and anticrossings. This system is chosen because detailed microscopic calculations are available for the same case [7]. Figure 5.4 compares the results of the two, but this requires some explanation. The microscopic calculations yield the amplitudes of the actual atomic displacements — in this case anions and cations alternately — while the phenomenological model yields continuos functions $\mathbf{u}(z) = \mathbf{u}_+(z) - \mathbf{u}_-(z)$.

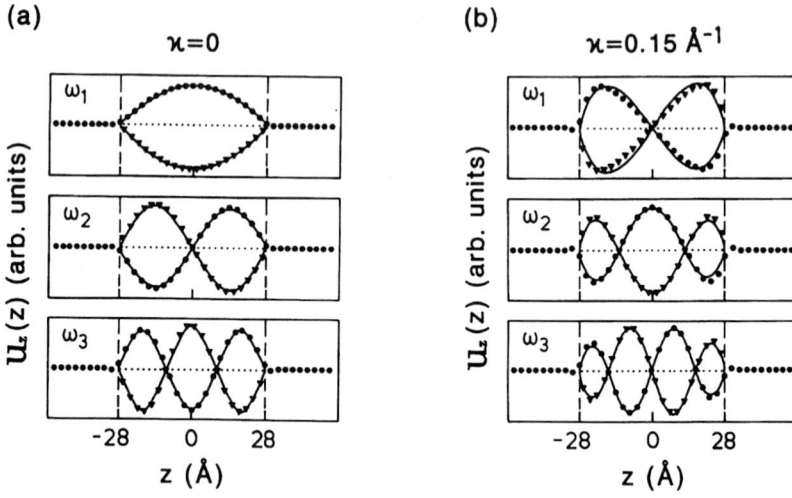

Figure 5.4 Spatial dependence of the *atomic* displacements *here and only here* denoted $u_z(z)$ (see text) for the GaAs quantum well of width $d = 5.6$ nm for $\kappa = 0$ (a) and $\kappa = 1.5 \times 10^7$ cm^{-1} (b). From top to bottom we have the $m = 1, 2, 3$ modes respectively. The labelling by increasing (decreasing) frequencies for each κ is obtained from the corresponding vertical cut in Figure 5.3. Note: There appears to be a confusion in the labelling of modes 1 and 3 in Figure 4 of Ref. [13]. Dots: anions. Triangles: cations.

What has been done in Figure 5.4 for the z component is to use the cation/anion inverse masses as weight factors to extract a continuos function which yields, within the continuum model, continuos 'atomic' amplitudes. At every successive discrete atomic position this gives the approximation yielded by the model to the actual atomic displacement of the corresponding atom. Here *and only here* this is denoted as $u_z(z)$ because it is compared with the results of the microscopic calculations [7], where the atomic displacements are denoted by this symbol. *Everywhere else* $\mathbf{u}(z)$, $u_y(z)$ and $u_z(z)$ denote *relative* displacements, as defined from the start in the phenomenological model. The atomic displacements obtained from this model are seen to be (Figure 5.4) in remarkable agreement with those of the microscopic calculation, even for not insignificant values of κ. This, together with a satisfactory account of the key experimental features [4, 8–13] lends credence to the basic correctness of the phenomenological model discussed in Chapters 3 and 4.

GaAs-$Al_xGa_{1-x}As$ heterostructures

Firstly, we consider a GaAs quantum well of width $d = 20$ Å with $Al_{0.9}Ga_{0.1}As$ outside and apply the *complete* matching boundary conditions following (3.44–3.46). Figure 5.5 shows the dispersion relations for the GaAs eigenmodes. Since these are different eigenmodes of the same eigenvalue problem we can label them by a discrete label $m = 1, 2, \ldots$ We ascribe $m = 1$ to the highest mode at $\kappa = 0$, so m grows on going downwards in frequency at $\kappa = 0$.

We have concentrated mainly on the range of κ from zero to values of order 10^7 cm^{-1}, which represent a very small fraction of the Brillouin zone and yet span the range of physical interest. The frequency eigenvalues are in fairly good agreement with experimental data [14], especially in their spacings, with

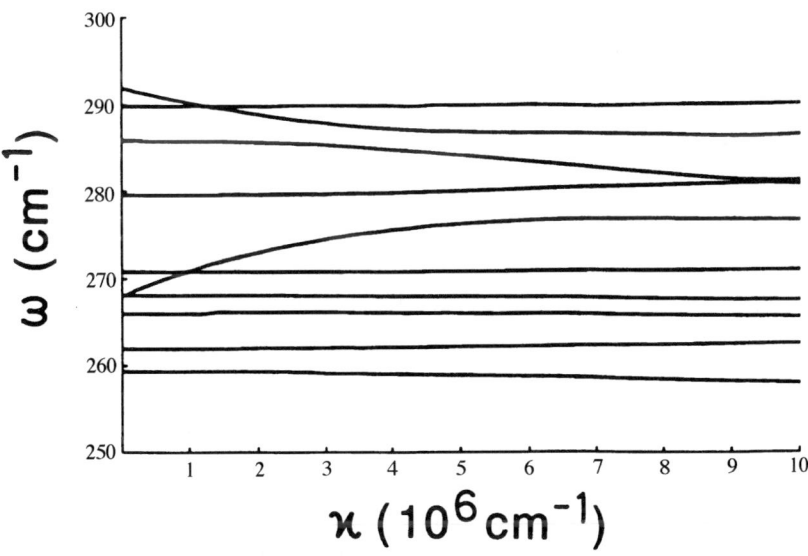

Figure 5.5 Dispersion relations for the GaAs eigenmodes of a 20 Å thick GaAs/$Al_{0.9}Ga_{0.1}As$ quantum well (frequency in cm^{-1}) and varying κ (in 10^6 cm^{-1}). Note the degeneracy of the modes $m = 6$ and 7 for $\kappa = 0$.

only a small systematic difference in the absolute value, which could be easily accounted for by a small inaccuracy in the determination of the well width. We also find a reasonable agreement between the results of complete matching and the infinite barrier model ($\mathbf{u} = 0$ at the interfaces). In any case the small numerical differences can be explained by the choice of the numerical values of the parameters involved.

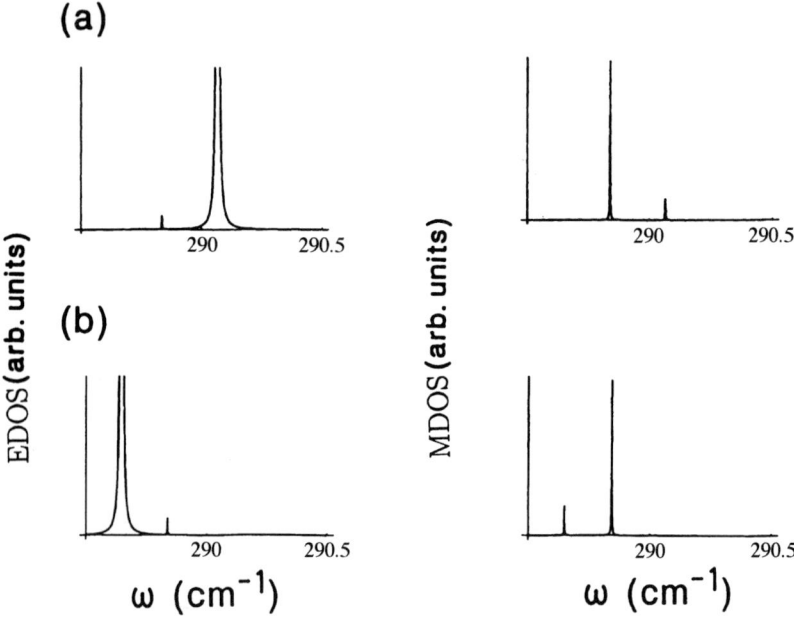

Figure 5.6 Electrical (left-hand side) and mechanical (right-hand side) spectral strengths for the upper and lower modes in the immediate neighbourhood of the crossover between modes $m = 1$ and $m = 2$ in Figure 5.5. (a) $\kappa = 1.1 \times 10^6$ cm^{-1} ($< \kappa_c$) and (b) $\kappa = 1.4 \times 10^6$ cm^{-1} ($> \kappa_c$). Spectral strengths in arbitrary units.

It is interesting to see the physical nature of the normal modes in this range. This may be done by separating out the contribution of the last diagonal element of the Green function $\mathbf{G}_s$ [15]. In Figure 5.6 the *electrical* and *mechanical* components of the density of states are shown for the first two modes for two values of κ on both sides of κ_c, the crossover value of κ at which modes 1 and 2 cross each other. This study indicates again that the more dispersive is the mode, the stronger its electrical character, i.e. the mode is mainly due to the electrical pole (section 4). On the other hand the non dispersive modes are mainly due to the mechanical poles, although both have, qualitatively speaking, a mixed character. Starting from $\kappa = 0$ there are two very dispersive modes which cut across the rest of the spectrum. The lower ($m = 6$) originates the second crossover with mode $m = 5$ for low κ. On looking at the entire system of eigenmode branches there are two, for every κ, which have predominantly electrical character. In the upper part of the spectrum this is $m = 1$. A similar situation holds about the lowest crossover, where $m = 6$ is the more dispersive.

Figure 5.7 shows the spatial dependence of u_z and φ for the first three modes when $\kappa = 2 \times 10^6$ cm$^{-1} > \kappa_c$. There is a nonvanishing amplitude u_y, which is nevertheless much smaller and is not displayed.

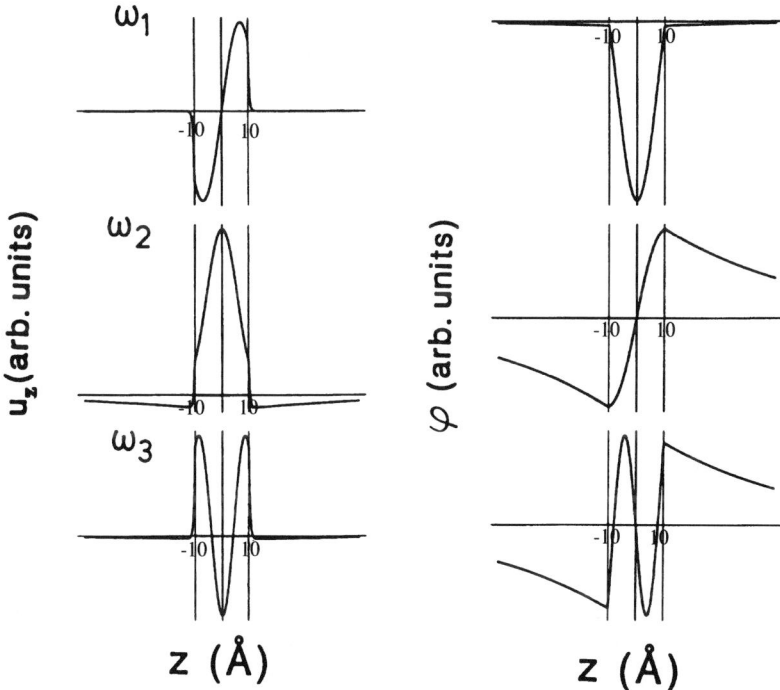

Figure 5.7 Spatial dependence of the mechanical (u_z: left hand side) and electrical (φ: right hand side) amplitudes of the quantum well of Figure 5.5 for $\kappa = 2 \times 10^6 cm^{-1}$, modes $m = 1, 2, 3$. The abscissa z is in Å. We stress that $u_z(z)$ denotes again the z component of $\mathbf{u} = \mathbf{u}_+ - \mathbf{u}_-$.

It is in order to remark again that while the physical system under study — the quantum well — is symmetric, the differential system (3.6–3.7) is not invariant under the reflection $z \to -z$. The vibration amplitude u_x is decoupled from the other amplitudes, thus only u_z, u_y and φ are related. Needless to say, u_y and u_z denote again Cartesian components of $\mathbf{u} = \mathbf{u}_+ - \mathbf{u}_-$. It follows from the field equations that $u_y(z)$ and $\varphi(z)$ have always the parity opposite to that of $u_z(z)$, a fact which also shows in Figure 5.7. Thus we can still compare the parities of different modes on the understanding that this refers only to one of the amplitudes. As indicated above we classify the states as even (even potential modes) or odd (odd potential modes) according to the parity of the potential $\varphi(z)$. Then the parity sequence for φ_1 to φ_5 is O, E, O, E, O when $\kappa < \kappa_c$ and it changes to E, O, O, E, O when $\kappa > \kappa_c$ because of the crossing of the levels $m = 1, 2$ at $\kappa = \kappa_c$.

The modes of the quantum well are no longer strictly longitudinal or transverse, as expected for $\kappa \neq 0$ when matching at interfaces is involved. However, explicit evaluation for some of them shows that $|\nabla \times \mathbf{u}|$ is still much smaller (greater) than $|\nabla \cdot \mathbf{u}|$, so these modes are still in practice quasi-longitudinal (quasi-transverse). Other modes show a strong longitudinal-transverse mixing; modes 5 and 6 are good examples of this mixing: on moving backwards from κ_c towards $\kappa \to 0$ we find that the mixture of longitudinal and transverse character is still significant, so that as soon as κ increases from zero, even while staying still below κ_c, mode 5 soon ceases to be quasi-longitudinal and mode 6 to be quasi-transverse.

Finally, the model also yields the transverse modes $m = 7, 8, \ldots$. These are strictly transverse *for all* κ, as they have only the amplitude u_x, which is factorised out in the full 4×4 determinant. These modes are purely mechanical and correspond to the shear horizontal wave one always encounters in the theory of elastic surface waves (see Chapter 4). The parity sequence of $u_x(z)$ for these modes is $E, O, E, \ldots$.

Some final comments on the use of continuum models to study problems involving interfaces and, in particular, heterostructures like quantum wells or superlattices are here in order. Firstly, when planar interfaces are involved we Fourier transform in 2D and introduce the corresponding wavevector κ. This means that we project the 3D band structure so we now label the normal modes by κ and ω. This is simply a formal representation for the matching calculation which comes later, but the bulk dispersion relation actually determines ω as a function of the 3D wavevector $\mathbf{k}$. Thus, in the isotropic model, for a given κ the frequency ω starts from $\omega^2 = \omega_\alpha^2 - \beta_\alpha^2 \kappa^2$ ($\alpha = L, T$) and then takes all the values from there down to $\omega^2 = \omega_\alpha^2 - \beta_\alpha^2 \kappa^2 - \beta_\alpha^2 k_z^2$, where k_z spans the range from zero to $k_z^2 = k^2 - \kappa^2$. Outside this range and for k_z real there are no allowed modes of the medium in question. These details must be kept in mind when analytic continuation arguments are used for k_z imaginary.

The other aspect of the problem is related to the count of the number of degrees of freedom, raised in section 2. Concerning the κ-dependence the argument is essentially the same, but a new situation arises related to the propagation in the direction perpendicular to the interfaces. Consider, for instance, the quantum well. Mathematically, the secular equation may yield many new branches $\omega(\kappa)$ and not all are physically meaningful. Besides the two interface modes, the number of meaningful confined modes must be determined in consonance with the number of atomic layers contained in the finite slab. It is easy to see, by concentrating on the starting values at $\kappa = 0$ where the propagation is perpendicular to the interfaces, that for each type of polarisation the 'higher modes' to be discarded are the ones with the lowest frequencies. If we consider a superlattice, then the count is more direct. It suffices to see how many distinct minibands must result from the folding given the thickness of the superlattice period.

Effects of crystal anisotropy

So far we have considered isotropic media, which is often sufficient in practice, depending on geometry and on the problem studied. However, there are anisotropic effects, for instance, when one considers different in-plane directions spanning the angle between equivalent high symmetry directions. The formulation is the same but the algebra is more complicated when in the same field equations the stress tensor σ includes anisotropy. The simplicity of (5.7–5.9) is then lost and finding a new basis of linearly independent solutions is then rather more involved. Furthermore, the matching boundary conditions are also more complicated.

An exact treatment of anisotropic media, including the new form of the matching boundary conditions, can be formulated and carried out in terms of the surface Green function matching analysis discussed in Chapter 4, as has been done in the case of elastic surface waves for hexagonal [16] and cubic [17, 18] media. This might be a useful procedure if one is interested, for instance, in spectral functions [18]. While no general rule can be established until each particular case has been examined, it has been pointed out [19] that for some problems the effects of crystal anisotropy in GaAs-based heterostructures are in practice small, although they should be experimentally observable. In this case the effects of crystal anisotropy can be treated as a perturbation which is why a perturbative method was presented in Chapter 3 for heterostructures involving cubic media. This is the case of the systems this book is mostly interested in, that is, GaAs-based heterostructures and the like. As stressed in Chapter 3, it is simpler to start from the matched system and then introduce the effects of anisotropy as a perturbation. This is particularly simple when the rigid wall approximation holds good, in which case the application of the method outlined in Chapter 3 is straightforward [19].

Since for GaAs/AlAs interfaces the mechanically rigid wall model is an excellent approximation, then formula (3.89) allows us to calculate the new eigenvalues of the normal modes and, by subtracting from these the values obtained for the isotropic case, one can in this sense estimate the correction to ω^2. The calculation requires evaluation of the normalisation integrals for the vibrational amplitudes **u**.

Of course the norm of **u** is arbitrary and conventional — recall the discussion in section 3 — and indeed formula (3.89) shows that this cancels out of the formula for $\Delta\omega^2$, but the integrals must be evaluated before the final result is obtained.

This procedure was used in Ref. [19] to study $\Delta\omega^2$ in the different types of modes of the GaAs type for a 46 Å wide [001]-grown AlAs/GaAs/AlAs quantum well. The simplest case is that of the transverse horizontal mode with amplitude u_x which is also decoupled from the rest when the in-plane propagation vector is in the (010) direction. The final result for $\Delta\omega^2$ to first order in the expansion parameter λ measuring the strength of the anisotropy is a very simple formula, namely

$$\Delta\omega_n^2 = \frac{(\beta_c^2 - 2\beta_T^2)}{2}\left(\kappa^2 + \frac{n^2\pi^2}{d^2}\right), \qquad (5.24)$$

which explicitly displays the proportionality to the anisotropy parameter $(\beta_c^2 - 2\beta_T^2)$.

Figure 5.8 shows $\Delta\omega^2$ for the other modes where the electrostatic field is coupled to a mixed vibration partaking in general — except for $\kappa = 0$ — in the nature of longitudinal and transverse, as has been discussed in Chapter 4 and in precedent sections of the present Chapter. The upper part of the figure gives the overall picture for the first four modes, which conveys the idea of the very complex pattern resulting from an initial pattern already loaded with the complexity of coupling, mixings, crossings and anticrossings. This overview spans the range of κ (denoted in the figure as k_y) from zero to 10^8 cm^{-1}, which clearly overstretches the long wave model. More significant is the lower part, up to $\kappa = 10^6$ cm^{-1}, where the long wave model is unquestionably acceptable. Some of the modes shown in the upper part are here shown. The significant feature is that while the vanishing of $\Delta\omega^2$ for $\kappa = 0$ appears to be a general rule, the exception is the mode which originates from the confined transverse $n = 1$ mode, for which there is a finite correction $\Delta\omega^2$ at $\kappa = 0$. This is in the lower right corner of the figure, corresponding to symmetric potential modes.

Superlattices

We now apply the approach developed above to the study of polar optical modes in a superlattice. The main feature for this type of structure is the long range nature of the electrostatic field which couples different constituent slabs even if

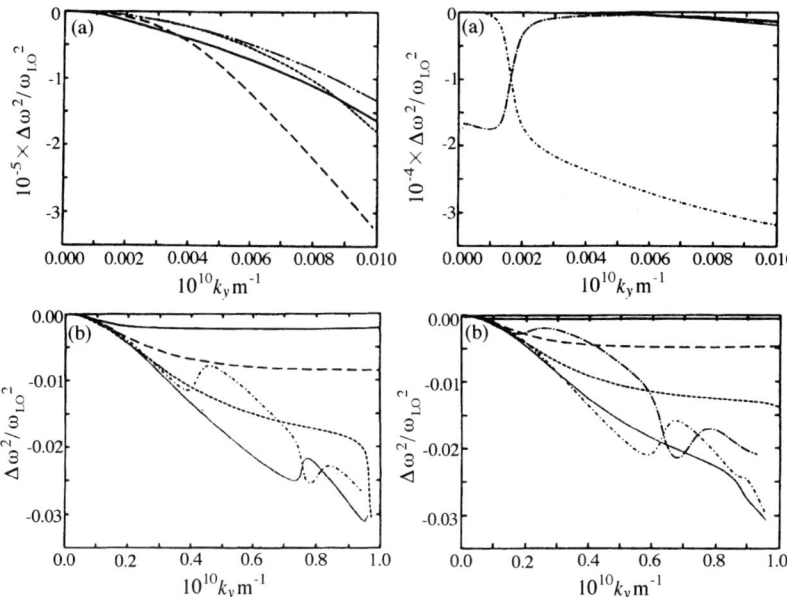

Figure 5.8 $\Delta\omega^2$ for the first few GaAs modes of a 46 Å wide AlAs/GaAs/AlAs quantum well. The in-plane value of κ is here k_y. The left hand side is for odd potential modes and the right hand side for even potential modes. The lower part shows some selected modes out of those appearing in the upper part.

the mechanical vibrations characteristic of each constituent material would not by themselves penetrate the adjoining slabs sufficiently to couple to the next slab of the same species. In order to concentrate on this feature we shall study the case of GaAs-AlAs superlattices, for which the assumption of infinitely rigid walls for the mechanical vibrations is a very good approximation. This allows for an analytical study which is then very transparent.

Given the simplicity of this analysis, it is interesting to extend the approach which has previously proved useful in the study of simpler structures since most experiments are actually carried out with superlattices, often arguing that since the mechanical vibrations of a given slab (say, GaAs) do not penetrate the adjoining (AlAs) slabs, one can treat, say, the normal modes of a GaAs slab as those of an isolated quantum well. We shall discuss the consequences of the long range electrostatic coupling even if the mechanical vibrations would by themselves remain uncoupled. Moreover, recent micro-Raman experiments [20, 21] bear out the distinct anisotropy of the *superlattice modes*, depending on the propagation direction relative to the in-plane and normal axes, a point we shall also examine.

For the quantum well structure we classify the solutions according to the parity of $\varphi(z)$, as stressed above, while for the superlattice we impose the Bloch condition

$$\mathbf{F}(z+nd) = e^{iqnd}\,\mathbf{F}(z), \tag{5.25}$$

$d = d_a + d_b$ being the superlattice period, d_a the GaAs layer width, d_b the AlAs layer width and q the wavevector associated with the superperiodicity. The same basis functions $\mathbf{F}_j(z)$ ((5.8) and (5.9)) can be used for the superlattice, only that now for z in a GaAs slab

$$\mathbf{F}(z) = \sum_{j=1}^{6} c_j \mathbf{F}_j(z) \tag{5.26}$$

with the material parameters of GaAs. Since we consider the GaAs modes, the solution in the AlAs slabs is only

$$\mathbf{F}(z) = e_1 \begin{bmatrix} 0 \\ 0 \\ 0 \\ e^{\kappa z} \end{bmatrix} + e_2 \begin{bmatrix} 0 \\ 0 \\ 0 \\ e^{-\kappa z} \end{bmatrix}. \tag{5.27}$$

The matching conditions are then the vanishing of u_y and u_z and the continuity of φ and D_z at the interfaces. This, together with the Bloch condition (5.25), results in superlattice modes consisting in general of combinations of all basis functions and then $\varphi(z)$ no longer has definite parity.

Between (5.26) and (5.27) we have eight unknown coefficients, which leads to an 8×8 secular determinant. This is considerably simplified if the basis of (5.27) is transformed in such a way that instead of e_1 and e_2 we have the two unknown coefficients h_1 and h_2, where

$$e_1 = \frac{(e^{-\kappa d_b/2} + e^{\kappa d_b/2 - iqd})h_1 - (e^{-\kappa d_b/2} - e^{\kappa d_b/2 - iqd})h_2}{4e^{-iqd}\sinh(\kappa d_b)}$$

$$e_2 = \frac{-(e^{\kappa d_b/2} + e^{-\kappa d_b/2 - iqd})h_1 + (e^{\kappa d_b/2} - e^{-\kappa d_b/2 - iqd})h_2}{4e^{-iqd}\sinh(\kappa d_b)} \tag{5.28}$$

Taking the unknowns in the order $c_1, c_2, c_3, h_1, c_4, c_5, c_6, h_2$ the secular equation takes the form

$$\begin{vmatrix} M_{OO} & M_{OE} \\ M_{EO} & M_{EE} \end{vmatrix} = 0, \tag{5.29}$$

where the submatrices are given by

$$M_{OO} = \begin{bmatrix} -x_\kappa \frac{\sin(x_L)}{x_L} & x_T \sin(x_T) & i\tilde{\alpha}_a x_\kappa \sinh(x_\kappa) & 0 \\ i\cos(x_L) & ix_\kappa \cos(x_T) & \tilde{\alpha}_a x_\kappa \cosh(x_\kappa) & 0 \\ i\frac{4\pi\alpha_a}{\epsilon_\infty^a}\frac{d_a}{2}\frac{\sin(x_L)}{x_L} & 0 & \frac{d_a}{2}\sinh(x_\kappa) & -1 \\ i4\pi\alpha_a \cos(x_L) & 0 & \epsilon_\infty^a x_\kappa \cosh(x_\kappa) & \epsilon_\infty^b \frac{2}{d_a} x_\kappa \frac{(\cosh(\kappa d_b)+\cos(qd))}{\sinh(\kappa d_b)} \end{bmatrix},$$

(5.30)

$$M_{OE} = \begin{bmatrix} 0 & 0 & 0 & 0 \\ 0 & 0 & 0 & 0 \\ 0 & 0 & 0 & 0 \\ 0 & 0 & 0 & -\epsilon_\infty^b \frac{2}{d_a} x_\kappa i \frac{\sin(qd)}{\sinh(\kappa d_b)} \end{bmatrix},$$

(5.31)

$$M_{EO} = \begin{bmatrix} 0 & 0 & 0 & 0 \\ 0 & 0 & 0 & 0 \\ 0 & 0 & 0 & 0 \\ 0 & 0 & 0 & \epsilon_\infty^b \frac{2}{d_a} x_\kappa i \frac{\sin(qd)}{\sinh(\kappa d_b)} \end{bmatrix},$$

(5.32)

$$M_{EE} = \begin{bmatrix} ix_\kappa \cos(x_L) & -i\cos(x_T) & i\tilde{\alpha}_a x_\kappa \cosh(x_\kappa) & 0 \\ -x_L \sin(x_L) & -x_\kappa \frac{\sin(x_T)}{x_T} & \tilde{\alpha}_a x_\kappa \sinh(x_\kappa) & 0 \\ \frac{4\pi\alpha_a}{\epsilon_\infty^a}\frac{d_a}{2}\cos(x_L) & 0 & \frac{d_a}{2}\cosh(x_\kappa) & -1 \\ -4\pi\alpha_a \sin(x_L) & 0 & \epsilon_\infty^a x_\kappa \sinh(x_\kappa) & \epsilon_\infty^b \frac{2}{d_a} x_\kappa \frac{(\cosh(\kappa d_b)-\cos(qd))}{\sinh(\kappa d_b)} \end{bmatrix},$$

(5.33)

with

$$x_L = \frac{k_{La} d_a}{2} \; ; \quad x_T = \frac{k_{Ta} d_a}{2} \; ; \\ x_\kappa = \frac{\kappa d_a}{2} \; ; \quad \tilde{\alpha}_a = \frac{\alpha_a}{\rho_a(\omega^2 - \omega_{La}^2)} \; .$$

(5.34)

The label a/b indicates GaAs/AlAs.

We first discuss the qualitative physical picture with special attention to two points, namely, (i) any possible symmetries and (ii) the similarities and differences between the superlattice and the quantum well modes.

The subdeterminants $|M_{OO}|$ and $|M_{EE}|$ are very similar to those yielding the odd potential modes and even potential modes, respectively, in the isolated quantum well, although both contain the (4,4) matrix element which displays distinct superlattice features. Thus they contain d_a and d_b, reflecting the coupling between a and b slabs. Moreover, in the order in which the sequence of unknown coefficients have been arranged these (4,4) elements pertain to the electrostatic part of the normal mode vibration, corresponding to the fact that the interslab coupling is electrostatic. This is more distinctly displayed in the off diagonal submatrices M_{OE} and M_{EO}. Firstly, they couple M_{OO} and M_{EE} and, secondly, they only contain the electrostatic element (4,4) with the b parameters, which describes the propagation of the electrostatic vibration through the b slabs. Of course, since in the successive slabs of the same material the mechanical and electrical vibrations are coupled, the electrostatic coupling induces a (smaller indirect) mechanical coupling similar to the one found in Ref. [22] for double barrier structures. In this case the coupling extends across the entire superlattice.

For given κ, the magnitude of the coupling terms depends on the thickness of the b slab. For $\kappa d_b \to \infty$ the off diagonal submatrices vanish, the total secular determinant is factorised as the product $|M_{OO}| \times |M_{EE}|$ and both these submatrices become identical to those of the isolated quantum well [8]. Thus, for given d_b the effects of electrostatic coupling between a slabs are significant in the range $\kappa < 1/d_b$, which is of course what one would expect. The results of (5.29) through (5.34) give the quantitative estimate of this effect.

A distinct feature of the superlattice modes is the dependence on q, the wavevector associated with the propagation in the growth direction. For arbitrary q the resulting $\varphi(z)$ has no definite parity, as stressed above, but for $q = 0, \pi/d$ the off diagonal matrices vanish and then $\varphi(z)$ has definite parity about the central slab where $z = 0$ is defined to be at its centre. It is interesting to study the long wave limit, to lowest order in κ, of these two cases.

For $q = 0$ we have

$$\|M_{OO}\| \simeq -ix_\kappa x_{Ta} \sin(x_{Ta})\epsilon_\infty^b \frac{d_a}{d_b} \times$$
$$\left[\left(1 - \frac{\epsilon_0^a}{\epsilon_\infty^a}\right) \frac{\omega_{Ta}^2}{\omega^2 - \omega_{La}^2} \left(\frac{\sin(x_{La})}{x_{La}} + \frac{d_b}{d_a} \frac{\epsilon_0^a}{\epsilon_\infty^a} \cos(x_{La})\right) + \left(1 + \frac{d_b}{d_a} \frac{\epsilon_\infty^a}{\epsilon_\infty^b}\right) \cos(x_{La})\right],$$

$$\|M_{EE}\| \simeq i\epsilon_\infty^a x_\kappa^2 x_{La} \sin(x_{La}) \times$$
$$\left[\left(\frac{4\pi\alpha\tilde{\alpha}}{\epsilon_\infty^a} - 1 - \frac{d_b}{d_a}\frac{\epsilon_\infty^b}{\epsilon_\infty^a}\right)\cos(x_{Ta}) - \frac{4\pi\alpha\tilde{\alpha}}{\epsilon_\infty^a}\frac{\sin(x_{Ta})}{x_{Ta}}\right].$$

(5.35)

Polar optical modes in layered structures

All the normal modes can then be classified in longitudinal and transverse.

The longitudinal odd potential modes tend to have the same eigenfrequencies as the corresponding modes for the isolated quantum well when $d_b \gg d_a$, while the transverse odd potential modes have always the same eigenfrequencies as in the isolated quantum well. The situation is the opposite with the even potential modes: Here, the longitudinal modes have always the same eigenfrequencies as in the isolated quantum well, while for the transverse modes this happens when $d_b \gg d_a$.

On the other hand, for $q = \pi/d$ we have

$$\|M_{OO}\| \simeq$$
$$-ix_\kappa x_{Ta} \sin(x_{Ta})\epsilon_\infty^b \left[1 + \left(1 - \frac{\epsilon_0^a}{\epsilon_\infty^a}\right) \frac{\omega_{Ta}^2}{\omega^2 - \omega_{La}^2}\right] \cos(x_{La}), \quad (5.36)$$

$$\|M_{EE}\| \simeq -i\frac{d_a}{d_b}\epsilon_\infty^b \cos(x_{Ta}) x_{La} \sin(x_{La}). \quad (5.37)$$

All normal modes can again be classified in longitudinal and transverse and in this case all of them — longitudinal or transverse and odd potential modes or even potential modes — have the same eigenfrequencies as in the isolated quantum well.

The case $\kappa = 0$ (whether $q = 0$ or π/d) requires some care. The determinant $|M_{OO}|$ appears then to have a root at $\omega = \omega_{Ta}$ while $|M_{EE}|$ has it at $\omega = \omega_{La}$. However, it is easily seen that both these roots have identically vanishing amplitudes so this is a spurious root to be discarded.

We recall that in an isolated quantum well all normal modes tend to be confined in the a slab except for two, which tend to accumulate their amplitude near the interfaces, hence the term *interface mode* for these two. However, this accumulation is imperceptible for very small κ. A physically more meaningful way of characterising these modes is to note that their spectral strength is essentially electrical rather than mechanical [8, 23], while the rest are essentially mechanical. An extreme case which bears this out is that of a single heterojunction, where the only matching solution — interface mode — is essentially of an electrical character (Chapter 4) and the electrostatic potential accumulates near $z = 0$ as $\exp(-\kappa|z|)$ [9]. The two interface modes of the quantum well result from the splitting of the degenerate modes at the two interfaces and retain their key physical feature — essentially electrical nature — even for $\kappa \simeq 0$, when the electrical potential is nearly constant as a function of z. The same basic features remain for the superlattice modes if the eigenfrequencies are studied as a function of κ for fixed q. However, the interest is in the simultaneous dependence of ω on κ and q, for which we shall presently show some results for a $(GaAs)_{12}$-$(AlAs)_{12}$ superlattice.

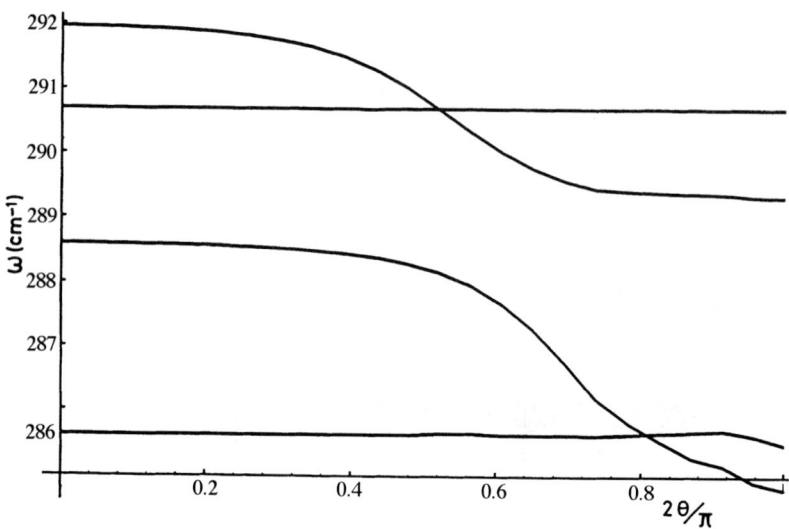

Figure 5.9 Anisotropy of the first four normal modes near the zone centre for a (GaAs)$_{12}$-(AlAs)$_{12}$ superlattice: ω vs. θ — see text.

Figure 5.9 displays the anisotropy of the long wave modes of a (GaAs)$_{12}$-(AlAs)$_{12}$ superlattice ($d_a = d_b = 34.32$ Å). These are states near the zone centre, corresponding to a wavelength $\lambda = 3500$ Å in a micro-Raman experiment [20, 21]. With a refractive index 3.5 this corresponds to a wavevector $K = 8.298 \times 10^5$ cm^{-1}. The κ and q components of the intervening normal modes are $\kappa = K \sin\theta$ and $q = K \cos\theta$ and the figure shows the eigenfrequencies of the normal modes as a function of θ for the said value of K. The crossings and anticrossings of the curves, depending on the symmetries — this is only in an approximate sense, since only states for $\kappa = 0$ and $q = 0, \pi/d$ have strict definite parity — are like those found in Ref. [24], though the theoretical analysis is very different.

Figure 5.10 is a 3D plot of ω as a function of κ and q for the same superlattice. The figure displays only the first two modes. Starting from $\kappa = 0$ (left hand side) and for given constant q, the first mode — from the top — is very dispersive and is the higher interface mode, as in the isolated quantum well [8, 23]. The second one corresponds to the second mode of the isolated quantum well [8, 23] and its frequency is nearly constant as a function of both κ and q. The "branches" ω vs. θ of Figure 5.9 result from cutting Figure 5.10 — and its continuation into lower frequency modes — by a small vertical cylinder, centred at the Γ point, of radius $K = 8.298 \times 10^5$ cm^{-1}.

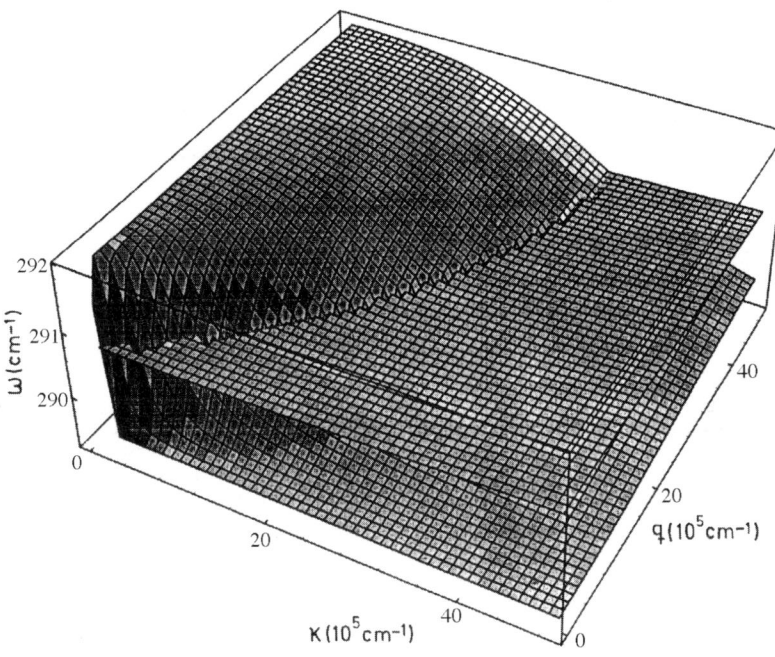

Figure 5.10 Threedimensional plot of ω (in cm^{-1}) as a function of κ and q for the first two normal modes of a (GaAs)$_{12}$-(AlAs)$_{12}$ superlattice. κ and q in 10^5 cm^{-1}. q is varied from 0 to π/d and κ is also varied between the same limits.

The rigid wall model for GaAs/AlAs heterostructures allows for a very transparent analytical study of the superlattice modes in which one can see the physical nature of the eigensolutions without any *ad hoc* assumption as well as pinpointing the effect of the long range electrostatic interaction which couples alternate slabs through the intermediate mechanically rigid layers. The superlattice spectrum consists of GaAs modes, on the one hand, and AlAs modes on the other. Here we have discussed only the former but of course the latter can be equally studied *mutatis mutandis*.

A similar but different calculation for GaAs/AlAs superlattices can be found in Ref. [26]. The calculation, which follows the general line of argument of Refs. [9] and [27], starts from a trial form of the solution written as a linear combination of longitudinal optical, transverse optical and interface modes. The results are very similar to the ones described here and also contain a mixing of interface and confined modes.

The difference with the present analysis is that no assumption is made here.

The problem is directly solved in terms of a basis and the results automatically contain all the key features. Besides displaying the role of the long range electrostatic coupling, the eigensolutions $\mathbf{u}_n(\mathbf{r})$ and $\varphi_n(\mathbf{r})$ are obtained in the basis formed by independent subspaces of transverse horizontal, odd potential and even potential modes. This provides a very convenient way of obtaining the field amplitudes $\mathbf{u}(\mathbf{r}, t)$ and $\varphi(\mathbf{r}, t)$ and hence the electron-phonon interaction Hamiltonian.

The same analysis can be made for any pair of constituent materials with sufficiently different normal frequencies that the assumption of mechanical rigidity is a good approximation. However, there are other cases, like GaAs-$Al_xGa_{1-x}As$ with x small, for which this does not hold. The formulation of the problem is then the same, but the solution would then be more complicated. In this case one can conveniently resort to the Green function technique (see Chapter 4) which provides a useful method both to solve the problem and to extract the desired physical information from the normal mode solutions. The physical picture would be on general lines the same but the matching rules then include also conditions of mechanical continuity [9, 23] and thus all frequencies — or most of them — propagate through all constituent slabs.

Double barrier structures

Consider the double barrier structures grown for the study of resonant tunnelling, where phonon replicas are observed in the tunnelling current [25] (GaAs - $Al_xGa_{1-x}As$ - GaAs - $Al_xGa_{1-x}As$ - GaAs). This indicates that there is a significant interaction of the tunnelling electrons with the phonons of the double barrier resonant tunnelling structure and therefore it is interesting to study the optical modes of such a system. Now, the mechanical amplitudes of the modes centred at the GaAs well decay very fast on penetrating the $Al_xGa_{1-x}As$ barriers, but the electrostatic potential does not, as seen in explicit calculations of the spatial dependence of amplitudes and spectral strengths for quantum wells (see section 5 and Ref. [22]) and single interfaces (see Chapter 4 and Ref. [10]). Since mechanical and electrical amplitudes are coupled, the normal modes of the GaAs well are ultimately coupled to the continuum of the bulk GaAs modes outside the barriers and this requires studying the modes of the entire double barrier structure, which involves matching at *four* interfaces.

The Surface Green Function Matching method introduced in Chapter 4 [15] is particularly efficient to treat an arbitrary number of interfaces [23]. It is easy to develop an algebra which effects the simultaneous projection at any number of surfaces so that matching is simultaneously effected in compact form at all the interfaces. This can be directly applied to the four interface problem involved in the double barrier structure but, since this is symmetric, it is interesting and convenient to use the method discussed in Chapter 4 for separating solutions of opposite parity. As a general remark we stress that the method explained in section 4 is valid not only to obtain the standing wave solutions of the double barrier, where the symmetry is more intuitively obvious, but also the directional propagating solution. The point is that the Green functions contain the information on the entire spectrum including propagating states and this is why the symmetrised Green functions yield the correct result even for the propagating solutions, when their directional nature would make it less intuitive. The details are explained in Ref. [28]. However while it is interesting to keep this in mind as a general observation, in this case we shall only discuss the quasistationary normal modes reflecting the effects of the long range electrostatic coupling across the barriers. In this case one needs only match at *two* interfaces instead of *four*. For a system of at least three coupled differential equations [27] this is a non negligible practical convenience. Another practical advantage will be seen presently.

We now describe the results of appliyng this analysis to the calculation of the spectrum of polar optical modes in the GaAs-based double barrier resonant tunnelling structures studied in Ref. [25]. As in all other cases, in the long wave phenomenological model used throughout the book the transverse horizontal vibration is factorised out and we are left with a κ-dependent Green function mixing u_y, u_z and φ which is a 3 × 3 matrix. The elements of this matrix for each constituent media were given in Chapter 4 and we recall that 1, 2 and 3 correspond to u_y, u_z and φ, in this order.

In keeping with the notation of Figure 4.4 the structure under study is $G_2 - G_1 - G_0 - G_1 - G_2$. Medium 2 is GaAs and medium 1 is the ternary compound $Al_xGa_{1-x}As$ with given fraction x of Al. Medium 0 is the same as medium 1 (GaAs) but on exploiting the symmetry the matching is effected only for the two interface system $G_{0p} - G_1 - G_2$ and G_{0p} is formed from G_0 — which by itself would be equal to G_2 — by using the analysis of section 4.

A comment is in order here. For large values of x, the Al fraction, the bottom of the conduction band of the ternary compound ceases to be at the Γ point, as it is in GaAs and when this happens this invalidates the widespread use of the one band effective mass model for the electronic states of the heterostructure. Now, the description in more elaborate models of electrons tunnelling through a double barrier structure would be quite cumbersome and so the tendency in general is to grow these heterostructures with smallish values of x. But then the rigid wall model for phonons ceases to be an acceptable assumption and full matching is then required.

In order to demonstrate the effects of electrostatic coupling through the barriers it will suffice to give the main results for just one case. Of the three double barrier resonant tunnelling structures studied in Ref. [10] we shall select the one having the widest barriers (d_b=111 Å) and largest composition

Table 5.1 Spectral strength (in brackets) of the main electrical peaks, in common arbitrary units. The asterisk (*) indicates that the calculation has been made for $\kappa=10^6$ cm^{-1}. For the rest $\kappa=10$ cm^{-1}. First row: Frequencies in units of cm^{-1}. Second row: Bulk values of ω_L. AlGaAs is the ternary compound with $x=0.4$. E_{QW}/O_{QW} denote the two interface (predominantly electrical) modes (even/odd) of the quantum well (QW). For the double barrier (DB) structure they appear at the same frequencies and two new interface modes, O'_{DB} and E'_{DB} appear very close to these.

ω	273.53	273.58	278.57*	292.22	292.35	292.37
ω_L		Al$_{0.4}$Ga$_{0.6}$As				GaAs
QW	$E_{QW}(1440)$		$E^*_{QW}(0.0138)$	$O_{QW}(0.01)$		
DB	$E_{DB}(0.238)$			$O_{DB}(0.0202)$		
DB		$O'_{DB}(0.0134)$			$E'_{DB}(8100)$	

difference ($x = 0.4$). This is the case when one would expect the weakest coupling effects. The well width is $d_b = 58$ Å and the other input parameters are taken from table 1.1.

In the following we refer always to GaAs-like modes which are the ones of interest for the phonon-assisted tunnelling [25] as the fraction of time spent by the tunnelling electrons in the AlGaAs barriers is comparatively negligible. After making full use of the formal symmetry analysis of section 4 we classify the modes in even/odd (E/O) by reference to the parity of the electrostatic potential in the *literal* sense that $\varphi_\kappa(z) = \pm\varphi_\kappa(-z)$. In this — literal — sense this is the same as the parity of u_y but opposite to that of u_z, as in the quantum well case.

In order to see the effects of long range electrostatic coupling between the central layer, the well region, and the external media across the barriers we compare the actual double barrier structure A-B-A-B-A with the corresponding well B-A-B in which the barriers extend to $\pm\infty$. We focus mainly on the long wave limit and study the electrical part of the density of states, which is the part of the total spectral strength which matters for the electron-phonon interaction. For a full study of the problem one would need the spatial dependence of the electrostatic potential $\varphi_\kappa(z)$ properly normalised [11]. This can be obtained if desired, but in order to demonstrate the effects of electrostatic coupling it suffices to note that for very small κ the predominantly electrical modes are the interface modes [10, 22], so we shall simply concentrate on the electrical spectral strength projected at the matching interfaces, which is sufficiently meaningful to demonstrate the changes in the spectrum of the well modes due to coupling across finite barriers.

We first consider the simple quantum well. With two matching interfaces

Polar optical modes in layered structures 109

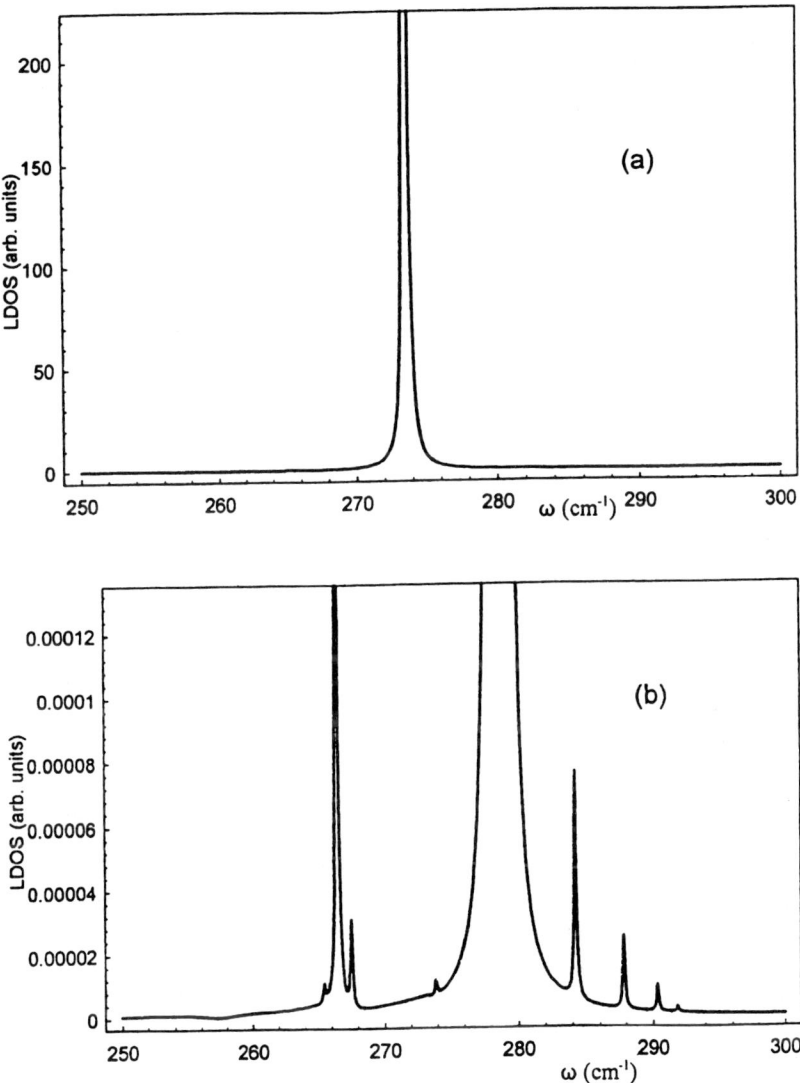

Figure 5.11 Electrical spectral strength for even potential modes in the simple quantum well described in the text. (a) Calculated for $\kappa = 10$ cm^{-1}. (b) For $\kappa = 10^6$ cm^{-1}. Spectral strength in arbitrary units, the same everywhere in this and the following figures. Peak intensities are given in Table 5.1 in the same units.

there are two interface modes, one odd (O_{QW}) and one even (E_{QW}), in order of decreasing frequencies. These appear in the electrical spectral strengths as peaks at the corresponding eigenfrequencies, given in Table 5.1 together with their peak intensities. The calculation was made for a very small $\kappa = 10$ cm^{-1}. The idea is to see what happens at the Γ point of the Brillouin Zone, but to do the calculation strictly for $\kappa = 0$ presents numerical complications. These are not impossible to overcome, but it is simpler to calculate for a very small value of κ, of the order of 10^{-7} times the wavevectors corresponding to the Brillouin Zone border. For all practical purposes this is effectively like sitting at the Γ point. It is striking that the strength of E_{QW} is *five* orders of magnitude higher than that of O_{QW}, a fact which can be understood by noting that E_{QW}, for $\kappa=10$ cm^{-1} is extremely close to the bulk resonant frequency ω_L for the barrier material (Al$_{0.4}$Ga$_{0.6}$As) outside the well. Figure 5.11 shows the electrical spectra of the even modes for $\kappa=10$ cm^{-1} (a) and for $\kappa=10^6$ cm^{-1} (b), which is still in the long wavelength range, two orders of magnitude smaller than a 'radius' of the Brillouin zone. The corresponding peak (E^*_{QW}) has moved upwards in frequency, away from ω_L (Al$_{0.4}$Ga$_{0.6}$As) and its strength (Table 5.1) has gone down by five orders of magnitude, now being comparable to the strength of O_{QW}.

Figure 5.12 compares the odd spectra for the quantum well and the double barrier structure. The eigenvalue of O_{QW}, again for $\kappa=10$ cm^{-1}, is quite insensitive to the change on going over from the simple quantum well to the double barrier structure, and the same holds for E_{QW}, but other features are rather more strongly affected. Thus, the strength of E_{DB}, fifth row in Table 5.1, goes down by four orders of magnitude with respect of that of E_{QW}. The mode still resonates with ω_L (Al$_{0.4}$Ga$_{0.6}$As) but outside the well there are only *finite* barriers of Al$_{0.4}$Ga$_{0.6}$As and the effect is much weaker. However the strength of E_{DB} is still one order of magnitude higher than that of O_{DB}, which remains essentially unchanged, although it increases by a factor of 2 because the spectra for the double barrier structure contain the contributions from the projection at *four* matching interfaces, instead of the *two* of the quantum well. Besides, two new modes (O'_{DB} and E'_{DB}) appear in the double barrier structure. The first one appears, accidentally, at the same frequency as E_{DB} and the second one very close to O_{DB}. The formal method for factorising separately the even and odd spectra is here very useful. Without this the problems of numerical resolution appearing in this case would be extremely cumbersome in practice. The mode E'_{DB} is now very close to the resonant frequency ω_L of the GaAs material of the well, where these modes mainly reside, and its strength goes up, again, to the order of magnitude of E_{QW}.

It is significant that the peak O'_{DB}, although also at the resonant frequency ω_L (Al$_{0.4}$Ga$_{0.6}$As) is one order of magnitude weaker than E_{DB} and of a strength similar to that of O_{DB}. A glimpse at a possible interpretation can be seen in

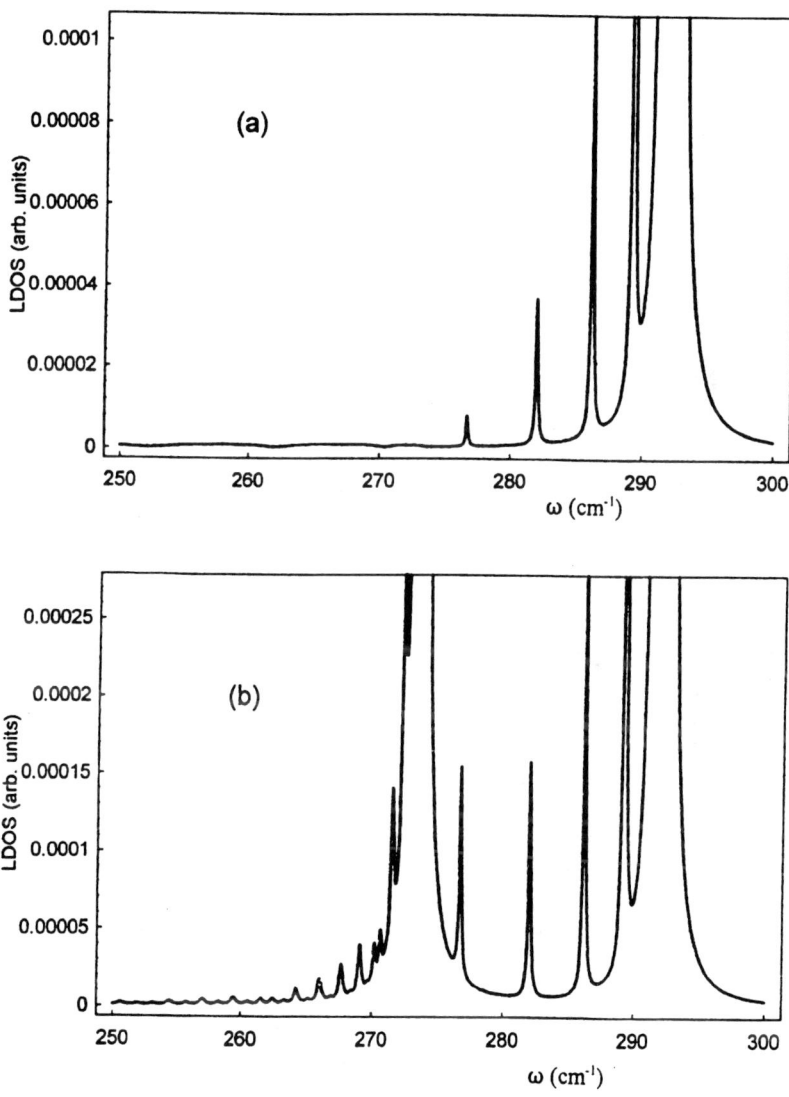

Figure 5.12 As in Figure 5.11, electrical spectrum of odd potential modes for $\kappa=10$ cm^{-1}. (a): For the simple quantum well. (b): For the double barrier resonant tunnelling structure.

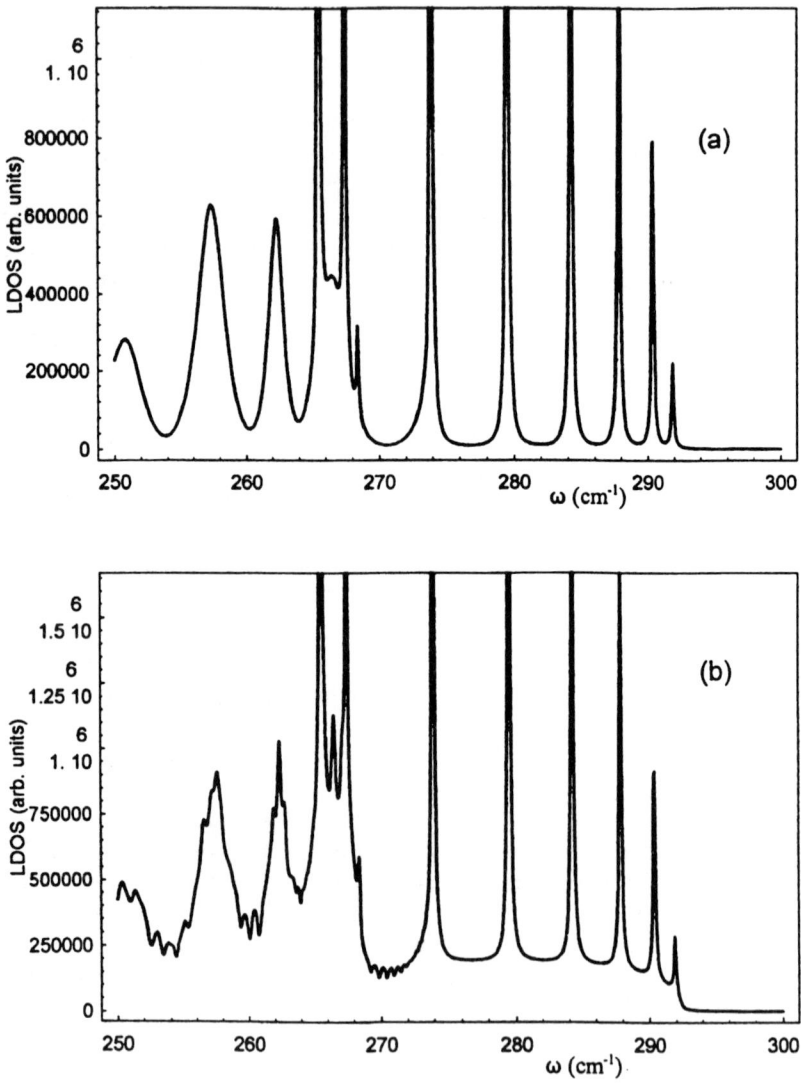

Figure 5.13 Mechanical spectrum of even potential modes for $\kappa = 10 \text{ cm}^{-1}$. (a): For the simple quantum well. (b): For the double barrier structure. Common arbitrary units for (a) and (b), unrelated to those of Figures 5.11 and 5.12.

Figure 5.13, which is also interesting to consider for the sake of completeness. This figure compares the mechanical spectral strength of the quantum well (a) with that of the double barrier structure (b), again for $\kappa=10$ cm^{-1} and in arbitrary common units. The numerical values of the *mechanical* spectral strength must not be related to those of the *electrical* spectrum, as they have different physical dimensions. What matters is the comparison between the two mechanical spectra of Figure 5.13. The point is that the polar modes under study have a mixed character and the electrical and mechanical vibrations are mutually coupled. With $x=0.4$ the mechanical barriers are already in practice totally rigid and all GaAs-like modes would be separately confined in the three disconnected GaAs regions of the structure with no direct coupling across the barriers, but the electrical vibrations are coupled and this drives an indirect coupling of the mechanical vibrations, as seen in Figure 5.13.b: The mechanical spectrum undergoes substantial changes on going from the quantum well to the double barrier structure which means that some spectral strength is transferred from electrical to mechanical. In this respect we note that there is a distinctly confined mode of essentially mechanical character at a frequency very close to 273.53 cm^{-1}. The question of why should this affect the odd electrical mode O'_{DB} rather than the even one E_{DB} could only be elucidated by a more detailed study, but it is clear that, irrespective of all the details the spectrum of the double barrier structure may exhibit quite substantial changes with respect to that of the simple quantum well.

As in other types of heterostructures, the electrical spectral strength is mostly in the interface modes as long as we stay in the very long wave limit. This no longer holds for larger values of κ [22] but since it is the long waves that essentially intervene in the electron-phonon interaction, this might be dominated by the interface modes.

Since the electrostatic field is long range, we may expect some coupling across the barriers and this is what the calculations probed by studying the GaAs-like modes of the heterostructure residing mainly in the GaAs well. Some interesting findings appear already in the simple quantum well. In this case the even potential mode resonates with the normal frequency ω_L of the ternary compound forming the barriers. Under these circumstances one could practically attribute all the strength of the electron-phonon coupling to just the even interface modes for $\kappa \cong 0$. A simplified calculation based on just this would probably give rather good results.

In the double barrier structure here discussed the results show that, even for barriers which are rather thick and mechanically very rigid, the effects of long range electrostatic coupling can be very strong. There is, to begin with, the simple fact that two new interface, predominantly electrical, modes appear and then there may be again drastic changes in the spectral strength of the main electrical peaks by several orders of magnitude. This might depend on the design parameters but in any case the lesson is that the electron-phonon interaction may differ quite substantially between the simple quantum well and the double barrier structure. For the sake of completeness it is relevant to note that the mechanical part of the spectral strength is also substantially affected due first to the electrostatic coupling across the barriers and then to the electromechanical coupling characterising the polar optical modes.

It is clear that the physical picture which emerges on studying the coupling across a symmetric double barrier structure also holds qualitatively for asymmetric structures.

Interfaces with no common ion

The vibrational properties of a single plane of B atoms embedded in an AC binary semiconductor (see Figure 5.14) are defined as planar vibrational modes or localised interface phonon modes. The first condition to support the planar vibrational modes is that vibrations involving the AB bond exhibit modes at

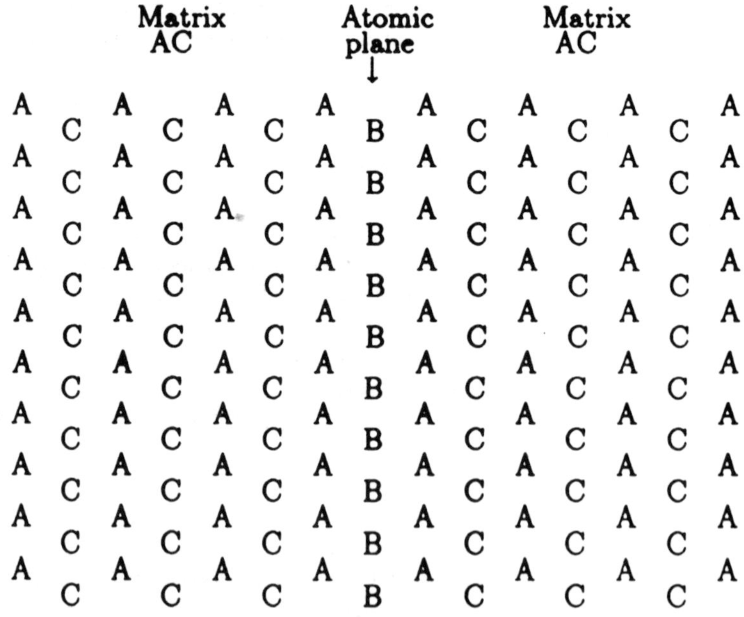

Figure 5.14 A plane of extraneous B atoms in an AC crystal matrix constitutes a planar defect capable of having localised planar vibrational modes.

frequencies which are forbidden in the host material; then the vibrational frequencies of an AB interface bond do not propagate into the bulk semiconductor and planar vibrational modes localised at the interface are formed. Experimentally this is practically realised as ultra-thin layer heterostructures and short period superlattices with no common ion, such as InAs/GaSb, [29, 30] and AlSb/InAs. [31–32]. Two distinct structures can be grown with either light (Ga-As or Al-As like bonds) or heavy (In-Sb bond) interfaces. The corresponding optical frequencies of GaAs, AlAs and InSb interface bonds present modes at energies above and below the bulk optical branches of the constituent materials. These systems have sharply localised quasi twodimensional interface phonons and their characteristics should be only expressed by interface bonds.

In heterostructures with a common ion (like GaAs/AlAs or GaAs/InAs) the experimental observation of localised vibrational modes is extremely difficult due to interaction between layers [33]. In the ultimate limit some structures have been grown with a single slab of a monoatomic layer of extraneous atoms embedded in an otherwise homogeneous material: GaAs/Al/GaAs [33, 34], GaSb/As/GaSb [32, 35], InAs/Al$_x$Ga$_{1-x}$/InAs [35]. Infrared absorption and Raman scattering experiments reveal the existence of these optical planar vibrational modes extremely confined in the neighbourhood of the interfaces. The study of planar vibrational modes has considerable impact on the characterisation of interface heterostructures and in the semiconductor growth technology.

This kind of localised optical mode was first reported by microscopic calculations in InAs/GaSb superlattices [36]. On the other hand, since the macroscopic phenomenological model works quite well for heterostructures in general, even for relatively thin layers [37, 38], it is interesting to study polar optical planar vibrational modes with this model, as one can then obtain to a reasonable approximation an analytical expression for the electrostatic potential of these modes and thus for the electron-phonon interaction Hamiltonian.

In Ref. [39] the macroscopic continuum approach, used to describe the optical phonon modes of GaAs/AlAs quantum wells and superlattices, has been extended to the treatment of long wavelength optical phonons in InAs/GaSb, AlSb/InAs, and related superlattices taking into account the inhomogeneity at interfaces. There, a 1D envelope function formalism is developed to describe the changes in reduced mass and optical force constant at the interface. This general case where the interface structure is important allows for an analysis of the strong interface effects that arise in systems with no common ion. Before studying planar vibrational modes with more complicated structures such as ultra-thin multiple quantum wells and short period superlattices with no common ion, it is convenient to have preliminary results for the ideal case of a single atomic plane of B atoms in a homogeneous semiconductor as the prototype of this kind of system.

From a phenomenological point of view the key feature is that new terms having the nature of extra forces appear at the interface. One is an inertial term, on account of mass differences which ultimately result in a local difference in the reduced mass density. This is described by a *surface excess* μ_0 of areal reduced mass density. The other one, associated with changes in force constants, results in a term having the nature of an extra stress at the interface. The simplest way is to introduce simply an extra force parameter γ_0 localised at the surface plane, so $\gamma_0 \delta(z)$ is the extra force term. Both these terms appear phenomenologically as extra normal 'forces' per unit area modifying the matching boundary conditions, which now have the form

$$\mathbf{u}(0+) = \mathbf{u}(0-), \tag{5.38}$$
$$\varphi(0+) = \varphi(0-), \tag{5.39}$$
$$D_z(0+) = D_z(0-), \tag{5.40}$$
$$\sigma_{3j}(0+) = \sigma_{3j}(0-) - (\gamma_0 - \mu_0 \omega^2) u_j(0); \quad j = 1, 2, 3. \tag{5.41}$$

The change appears in the modification of the continuity of the σ_{3j}. The form of the extra terms introduced in (5.41) is fully in line with abundant similar studies which have been carried out for various problems such as surface and interface effects in fluids [40] and solids [41]. The case of a planar defect has been similarly treated in the context of elastic waves [41, 42]. The general physical background for viscoelastic fluids can be found in Ref. [43] and for fluids and solids in Ref. [15]. The rest of the calculation is as usual. In each halfspace, and after elimination of non convergent functions at $z = \pm\infty$, the solution is taken as a linear combination of the basis given by Eqs. (5.8) and (5.9).

Transverse Horizontal Mode

In this case the conditions (5.38–5.41) are reduced to the continuity of u_x at $z = 0$ and

$$(\gamma_0 - \mu_0 \omega^2) u_x(0) + \sigma_{13}(0+) - \sigma_{13}(0-) = 0. \tag{5.42}$$

By application of the above conditions to a linear combination of basis functions a straightforward calculation leads to a transverse horizontal mode with frequency

$$\omega^2 = \omega_0^2 + \frac{2\rho^2 \beta_T^2}{\mu_0^2} \left[1 + \sqrt{1 + \frac{\mu_0^2}{\rho^2 \beta_T^2} \left[\omega_0^2 - \omega_t^2(\kappa) \right]} \right], \tag{5.43}$$

where $\omega_0 = \sqrt{\gamma_0/\mu_0}$. We define $\kappa_{L,T} = \sqrt{\kappa^2 - (\omega_{L,T}^2 - \omega^2)/\beta_{L,T}^2}$ and then the corresponding eigenvector has the form

$$u_x(z) = \sqrt{\frac{\kappa_T}{\rho}} e^{-\kappa_T |z|}, \qquad (5.44)$$

where κ_T is evaluated at the eigenfrequency ω. In Figure 5.15.a the amplitude $u_x(z)$ is shown as a function of z for $\kappa = 0$. For Al-plane atoms in a GaAs-matrix, $\beta_T = 3.3 \times 10^5$ cm/s [38] and $\omega_0 = 358$ cm^{-1}, the amplitude (5.44) decays by a factor $1/e$ for $z \approx 0.5\, a$.

Coupled solutions

In terms of the 3–component field

$$\mathbf{F} = \begin{pmatrix} u_y \\ u_z \\ \varphi \end{pmatrix} \qquad (5.45)$$

the space of solutions has the following symmetry property:

$$\mathbf{F}_\pm(z) = \pm \begin{pmatrix} 1 & 0 & 0 \\ 0 & -1 & 0 \\ 0 & 0 & 1 \end{pmatrix} \mathbf{F}_\pm(-z), \qquad (5.46)$$

where $\mathbf{F}_\pm$ is a linear combination of basis functions given in (5.8)–(5.9). From Eq. (5.46) and the matching conditions (5.38)–(5.41) we obtain an eigenstate with frequency $\omega_0 = \sqrt{\gamma_0/\mu_0}$ and eigensolution given, for $z < 0$, by

$$\mathbf{F}_+(z) = \\ A_+ \left[\begin{pmatrix} i\kappa^2 \\ \kappa_L \kappa \\ 4\pi \alpha \kappa/\epsilon_\infty \end{pmatrix} e^{\kappa_L z} - \kappa_L \begin{pmatrix} i\kappa \Omega_1 \\ \kappa \Omega_1 \\ 4\pi \alpha/\epsilon_\infty \end{pmatrix} e^{\kappa z} + i\kappa_L \Omega_2 \begin{pmatrix} -\kappa_T \\ i\kappa \\ 0 \end{pmatrix} e^{\kappa_T z} \right],$$

(5.47)

where

$$\Omega_1 = \frac{\omega_L^2 - \omega_T^2}{\omega_0^2 - \omega_T^2}, \qquad \Omega_2 = \frac{\omega_0^2 - \omega_L^2}{\omega_0^2 - \omega_T^2}, \qquad (5.48)$$

$$A_+ = \frac{1}{\sqrt{\rho}} \left[\frac{\kappa^4}{\kappa_L} - 3\kappa^2 \kappa_L + 2\kappa \kappa_L^2 \Omega_1^2 \right.$$
$$\left. + \kappa_L^2 \left(\frac{\kappa^2}{\kappa_T} + \kappa_T \right) \Omega_2^2 + 4\kappa \kappa_L^2 \Omega_1 \Omega_2 \right]^{-1/2}$$

(5.49)

and κ_T and κ_L are evaluated at $\omega = \omega_0$. The normalisation constant A_+ is obtained by following the procedure indicated in section 3. Figures 5.15.b-5.15.c present the mechanical amplitudes $u_y(z)$, $u_z(z)$ as functions of z, for $\kappa = 10^6$ cm^{-1}, $\beta_L = 3.3 \times 10^5$ cm/s [38] and $\omega_0 = 358$ cm^{-1}. We note that

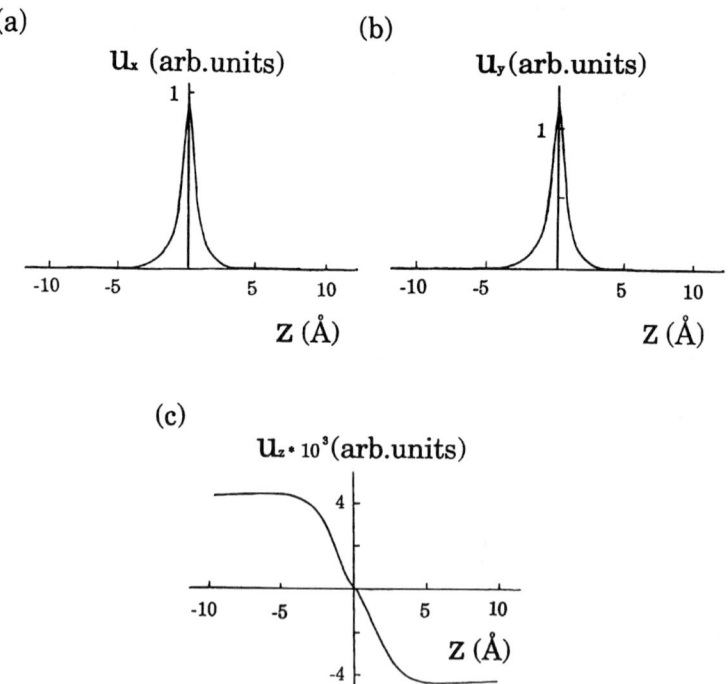

Figure 5.15 a) Mechanical amplitude $u_x(z)$ of the transverse horizontal mode (see Eq. (5.44)) for $\kappa = 0$ and $\omega_0 = 358$ cm^{-1}. Mechanical amplitude of the coupled mode for $\kappa = 10^6$ cm^{-1} and $\omega_0 = 358$ cm^{-1}. b) $u_y(z)$; c)$u_z(z)$. Common arbitrary units for b) and c), unrelated to Figure a). Values of GaAs parameters are used following table 1.1.

the vibration amplitude $u_z(z)$ vanishes at the interface while $u_y(z)$ presents a cusp at $z = 0$. When the characteristic frequency of the interface ω_0 is forbidden in the host material κ_L and κ_T are larger than the in-plane phonon wave vectors and then the κ-exponential in (5.48) dominates at long z-distance. An explicit calculation shows that $\nabla \times \mathbf{u} \neq 0$ while $\nabla \cdot \mathbf{u} = 0$ at $\kappa = 0$. Hence, this mode is quasi-transverse for low values of κ. The ratio $T = |\nabla \cdot \mathbf{u}|/|\nabla \times \mathbf{u}|$ evaluated at $y = 0$, $z = 0$ presents a linear dependence on κ (for $0 < \kappa < 4 \times 10^7$ cm^{-1}) showing that for $\kappa \neq 0$ the eigensolution (5.48) is no longer a purely transverse mode but a mixed one. A direct evaluation for $\kappa = 4 \times 10^6$ cm^{-1} yields $T = 0.4$, which indicates the mixed longitudinal and transverse character of this mode.

Since the transverse horizontal mode has no associated electrostatic potential, this solution is not expected to be optically active while the dipole operator responsible for infrared absorption is non zero. That is, the uncoupled transverse horizontal phonon is ir-active. Ref. [33] describes a study of the transverse horizontal mode in a single slab of Al atoms substituting for a Ga monoatomic layer in GaAs. Using infrared absorption spectroscopy these authors observed a peak at 358 cm^{-1} related to transverse optical phonons due to a two-dimensionally localised vibrational mode. The 358 cm^{-1} value is lower than the $\omega_T = 361$ cm^{-1} for AlAs bulk semiconductor. The observed absorption peak is a proper Al-like planar vibrational mode in a GaAs-matrix and is in correspondence with the model predictions here presented.

The second solution, with eigenfrequency $\omega_0 = \sqrt{\gamma_0/\mu_0}$, has a mixed character involving the mechanical amplitude and electrostatic potential and should be Raman active. The Raman transition operator for the dipole-allowed scattering at $\kappa \neq 0$ predicts a vibrational mode limited to a scattered peak at $\omega = \omega_0$. The coupled planar vibrational modes given by (5.48) must be observed by Raman scattering in the $Z(X,Y)\bar{Z}$ configuration (X,Y and Z stand for the direction [100], [010], and [001], respectively) with fixed frequency ω_0 only related to interface atoms. The excitation of Al-like planar vibrational modes in InAs has been reported by Raman scattering in [35]. The measured value of ω_0 is 349 cm^{-1} which, from $\omega_0 = \sqrt{\gamma_0/\mu_0}$, determines a zone-centre force of 3.64 N/m for the AlAs plane in an InAs matrix semiconductor. Raman scattering spectra from a Ga monolayer embedded in InAs and a plane of As atoms substituting for Sb plane in GaSb have been reported by Shanabrook and Bennett [33, 35]. The GaAs-like planar vibrational mode is observed strongly in the $Z(X,Y)\bar{Z}$ scattering geometry in both systems (InAs/Ga/InAs and GaSb/As/GaSb) at the same frequency value $\omega_0 = 254$ cm^{-1} giving a force constant $\gamma_0 = 3.5$ N/m.

The phenomenological model can thus give a satisfactory account of all essential features of the localised modes of extraneous atomic planes inserted in a matrix semiconductor, while yielding analytical solutions from which the electron-phonon Hamiltonian can be obtained. Furthermore, by fitting to experimental data one can estimate the zone-centre force constant for Al-like planar vibrational modes in a InAs matrix and GaAs plane in GaSb or InAs binary semiconductors. Thus, by measuring the phonon frequencies by infrared absorption and Raman scattering, one can estimate interface or matrix environment semiconductor parameters.

References

1. S. Adachi, J. Appl. Phys. **58**, R1 (1985).

2. Z.P. Wang, D.S. Jiang and K. Ploog, Solid State Commun. **65**, 661 (1988).

3. M. Babiker, J. Phys. C: Solid State Physics **19**, 683 (1986).

4. C. Trallero-Giner and F. Comas, Phys. Rev. **B37**, 4583 (1988).

5. E.P. Pokatilov and S.E. Beril, Phys. Stat. Sol. (b) **118**, 567 (1983).

6. H. Gerecke and P. Bechstedt, Phys. Rev. **B43**, 7053 (1991).

7. H. Rücker, E. Molinari and P. Lugli, Phys. Rev. **B44**, 3463 (1991).

8. F. Comas, R. Pérez-Alvarez, C. Trallero-Giner and M. Cardona, Superlattices and Microstructures **14**, 95 (1993).

9. F. García-Moliner, in 'Phonons in semiconductor nanoestructures', edited by J.P. Leburton, J. Pascual and C. Sotomayor Torres, Kluwer Academic Publs., Dordrecht (1993).

10. A. Chubykalo, V.R. Velasco and F. García-Moliner, Surf. Sci. **319**, 184 (1994).

11. C. Trallero-Giner, F. García-Moliner and F. Comas, Phys. Rev. **B50**, 1755 (1994).

12. R. Pérez-Alvarez, V.R. Velasco and F. García-Moliner, Physica Scripta **51**, 526 (1995).

13. E. Molinari, S. Baroni, P. Giannozzi and S. de Gironcoli, Phys. Rev. **B45**, 4280 (1992).

14. A.K. Sood, J. Menéndez, M. Cardona and K. Ploog, Phys. Rev. Lett. **54**, 2111 (1985).

15. F. García-Moliner, *Theory of Simple and Multiple Interfaces*, World Scientific, Singapore (1992), p.421.

16. V.R. Velasco and F. García-Moliner, Surf. Sci. **83**, 376 (1979).

17. V.R. Velasco and F. García-Moliner, Solid St. Commun. **33**, 1 (1980).

18. V.R. Velasco and F. García-Moliner, J. Phys. C: Solid St. Phys. **13**, 2237 (1980).

19. M.P. Chamberlain, C. Trallero-Giner and M. Cardona, Phys. Rev. **B50**, 1611 (1994).

20. M. Haines and G. Scarmacio, in 'Phonons in semiconductor nanoestructures', edited by J.P. Leburton, J. Pascual and C. Sotomayor Torres, Kluwer Academic Publs., Dordrecht (1993).

21. G. Scarmacio, M. Haines, G. Abstreiter, E. Molinari, S. Baroni, A. Fischer and K. Ploog, Phys. Rev. **B47**, 1483 (1993).

22. R. Pérez-Alvarez, F. García-Moliner, V. R. Velasco, C. Trallero-Giner, Phys. Rev. **B48**, 5672 (1993).

23. R. Pérez-Alvarez, F. García-Moliner, V.R. Velasco and C. Trallero-Giner, J. Phys.: Condensed Matter **5**, 5389 (1993).

24. N.C. Constantinou, O. Al-Dossary and B.K. Ridley, Solid State Commun. **86**, 191 (1993).

25. N. Zou, J. Rammer and K.A. Chao, Int. J. Modern Phys. **7**, 3449 (1993).

26. M.P. Chamberlain, M. Cardona and B.K. Ridley, Phys. Rev. **B48**, 14356 (1993).

27. C. Trallero-Giner, F. García-Moliner, V.R. Velasco, and M. Cardona, Phys. Rev. **B45**, 11944 (1992). R. Pérez-Alvarez, F. García-Moliner, V.R. Velasco, and C. Trallero-Giner, J. Phys.: Condensed Matter **5**, 5389 (1993).

28. F. García-Moliner, R. Pérez-Alvarez, H. Rodríguez-Coppola and V.R. Velasco, J. Phys. A: Math. Gen. **23**, 1405 (1990).

29. I. Lee and C.Y. Fong, Phys. Rev. **B44**, 6270 (1991).

30. S.G. Lyapin, P.C. Klipstein, N.J. Mason, and P.J. Walker, in *Proccedings of the 22nd International Conference on the Physics of Semiconductors*, Vancouver, 1994, edited by D.J. Lockwood (World Scientific, Singapore, 1995), p.915.

31. I. Sela, C.R. Bolognesi, L.A. Samoska, and H. Kroemer, Appl. Phys. Lett. **60**, 3283 (1992).

32. B.V. Shanabrook and B.R. Bennett, Phys. Rev. **B50**, 1695 (1994).

33. H. Ono and T. Baba, Phys. Rev. **B44**, 12908 (1991).

34. H. Tanino and Sh. Amano, Surf. Sci. **267**, 422 (1992).

35. B.V. Shanabrook and B.R. Bennett, in *Proceedings of the 22nd International Conference on the Physics of Semiconductor*, Vancouver, 1994, edited by D.J. Lockwood, World Scientific, Singapore (1995), p. 955.

36. A. Fasolino, E. Molinari, and J.C. Maan, Superlattices and Microstructures **3**, 117 (1987).

37. M. Cardona, Superlattices and Microstructures **7**, 183 (1990).

38. F. Comas, R. Pérez-Alvarez, C. Trallero-Giner, and M. Cardona, Superlattices and Microstructures **14**, 95 (1993).

39. B.A. Foreman, Phys. Rev. **B52**, 12260 (1995).

40. E. H. Lucassen-Reynders and J. Lucassen, Advan. Colloid Interface Sci. **2**, 347 (1969).

41. V. R. Velasco and F. García-Moliner, Physica Scripta **20**, 111 (1979).

42. A. M. Kosevich and V. I. Khokhlov, Soviet Phys. Solid State **10**, 39 (1968).

43. L. D. Landau and E. M. Lifshitz, *Fluid Mechanic*, Pergamos Press NY (1959).

CHAPTER SIX

Quasi-1D semiconductor nanostructures

The auxiliary functions Ψ (scalar) and $\mathbf{\Gamma}$ (vector) introduced in Chapter 3 provide a general method of solution of the field equations which holds for any geometry. So the method of solution thus obtained can be used to study interfaces of any shape, provided it is tractable in practice. In this chapter we discuss the cylindrical case and apply this to the study of quantum wires.

Nowadays several advanced technologies, such as molecular-beam epitaxy and selective ion implantation, permit the growth of semiconductor microcrystals with quasi-1D properties (quantum well wires) [1–8]. Several phenomenological theories with various degrees of approximation have been applied also to the investigation of the polar oscillations in quantum well wires and freestanding wires. The long wave polar optical modes of a quantum well wire have been studied [1, 9–20] in the framework of either a hydrodynamic or a dielectric continuum model. The latter has been applied to a quantum well wire with square cross section in Refs. [9, 11, 19]

A critical analysis of the results of these studies has been given in [6, 16, 17], where also interface modes are analysed in different geometries and with different cross sections. The hydrodynamical model has been used [9, 17, 18] to study the optical modes in cylindrical wires. The optical oscillations were always assumed to be either purely longitudinal or transverse, and the coupling between the electrostatic potential associated with the longitudinal optical oscillations and the mechanical vibrations not taken into account. As explained in Chapters 3 and 4 this model gives an incorrect formulation of the interface modes and cannot reproduce the strong coupling between longitudinal optical and transverse optical phonons predicted by microscopic calculations [20, 21]. The physical arguments given in Chapters 3 and 4 are general and independent of the geometry, dimensionality and other aspects of the system or structure under study. Optical modes in cylindrical quantum wires have also been studied by a sort of *ad hoc* extension of the dielectric model in which lattice dynamics is used and the condition that the electrostatic potential must vanish at the surface of a rectangular quantum wire is imposed [17]. The differences between such kinds of dielectric continuum approach and microscopic calculations were discussed in [21]. Microscopic calculations of the vibrational frequencies for rectangular quantum well wires embedded in AlAs have been reported in [20–23]. The anisotropic character of the phonon dispersion in that kind of structure was analysed in [21] and it is in accordance with similar calculations in quantum wells. The GaAs and AlAs confined and interface modes are studied in [21] in terms of a lattice dynamics calculation.

Experimental evidence of surface modes has been reported in [1], where a Raman scattering measurement has been carried out in GaAs cylindrical wires of 300 Å radius. Evidence of confined modes in cylindrical quantum well wires has been presented in [24].

In this chapter we shall continue with the application of the model developed in Chapter 3 to quantum well wires and free standing wires. We recall that the main points of this treatment are:

1. To solve the system of coupled differential equations (3.28) and (3.29) for the vibration amplitude **u** and the electrostatic potential φ, respectively;

2. To apply the matching boundary conditions at the interfaces in close consistency with both the differential equations of the treatment and the physical principles involved (Eqs. (3.44)-(3.46));

3. In general, to study coupled oscillation modes having a mixed character.

We chose a cylindrical geometry with circular cross-section of radius r_0. We then discuss the nature of the oscillation modes with detailed reference to the dispersion relation curves, the oscillation amplitudes and the electrostatic potential. Moreover, a derivation of the Fröhlich-like electron-phonon interaction Hamiltonian will also be presented.

The cylinder cross-section is assumed to be circular with z axis along the centre of the cylinder axis of infinite length. We have a medium A inside ($r < r_0$) and medium B outside ($r > r_0$). We use cylindrical coordinates (r, θ, z), require regularity of the solutions at $r = 0$ and $r \longrightarrow \infty$ and impose matching boundary conditions at $r = r_0$. In order to solve the problem it is necessary to obtain the general solutions of Helmholtz's equation for Ψ (3.36) and Γ (3.32) and of the equation for φ_H (3.39) within each medium and to match them at the interface. Having obtained solutions for Γ, Ψ and φ_H, the functions **u** and φ can be obtained by means of equations (3.40) and (3.38), respectively. Thus we obtain general analytical solutions for the coupled fields **u** and φ. Instead of giving the most general solution we shall restrict ourselves to the investigation of the important case of oscillations perpendicular to the wire axis, with wavevector, $k_z = 0$. Then, $u_z = 0$ and the solutions do not depend on z. We are thus considering just a particular case which, however, suffices to give an insight into the nature of the cylinder normal modes. The results presently to be described could not be used for a direct evaluation of physical processes such as scattering rate or free carrier absorption, where $k_z \neq 0$. Nevertheless, this case entails a direct interest for the study of one phonon Raman scattering for a backscattering configuration along the z-axis [25]. For $u_z = 0$ we can take

$$\mathbf{\Gamma} = \Gamma_z(r,\theta)\mathbf{e}_z \tag{6.1}$$

and

$$\Psi = \Psi(r,\theta), \tag{6.2}$$

where $\mathbf{e}_z$ is the unit vector along the z axis. Thus, $\mathbf{u}$ is of the form:

$$\mathbf{u} = u_r(r,\theta)\mathbf{e}_r + u_\theta(r,\theta)\mathbf{e}_\theta , \tag{6.3}$$

where $\mathbf{e}_r$ and $\mathbf{e}_\theta$ are the corresponding unit vectors in cylindrical coordinates. It can be easily seen from Eq. (3.40) and Eqs. (6.1–6.3) that:

$$u_r = -\frac{\partial}{\partial r}\left(\frac{\alpha}{\rho \beta_T^2 Q_T^2}\varphi_H + \frac{\Psi}{Q_L^2}\right) + \frac{1}{Q_T^2}\frac{1}{r}\frac{\partial \Gamma_z}{\partial \theta}, \tag{6.4}$$

$$u_\theta = -\frac{1}{r}\frac{\partial}{\partial \theta}\left(\frac{\alpha}{\rho \beta_T^2 Q_T^2}\varphi_H + \frac{\Psi}{Q_L^2}\right) - \frac{1}{Q_T^2}\frac{\partial \Gamma_z}{\partial r}, \tag{6.5}$$

while Γ_z and Ψ satisfy the following equations:

$$\frac{1}{r}\frac{\partial}{\partial r}\left(r\frac{\partial \Gamma_z}{\partial r}\right) + \frac{1}{r^2}\frac{\partial^2 \Gamma_z}{\partial \theta^2} + Q_T^2 \Gamma_z = 0, \tag{6.6}$$

$$\frac{1}{r}\frac{\partial}{\partial r}\left(r\frac{\partial \Psi}{\partial r}\right) + \frac{1}{r^2}\frac{\partial^2 \Psi}{\partial \theta^2} + Q_L^2 \Psi = 0 . \tag{6.7}$$

We require the solutions to be regular in $r = 0$ and in $r \longrightarrow \infty$. It is straightforward to see that a general analytical basis for the solution space is:

$$\begin{bmatrix} \frac{in}{r} f_n(Q_T r) \\ -Q_T f_n'(Q_T r) \\ 0 \end{bmatrix} e^{in\theta} \quad ; \quad \begin{bmatrix} Q_L f_n'(Q_L r) \\ (\frac{in}{r}) f_n(Q_L r) \\ (\frac{4\pi\alpha}{\epsilon_\infty}) f_n(Q_L r) \end{bmatrix} e^{in\theta} \tag{6.8}$$

and

$$\begin{bmatrix} [\frac{n\alpha}{\rho(\omega^2-\omega_T^2)}]r^{n-1} \\ [\frac{in\alpha}{\rho(\omega^2-\omega_T^2)}]r^{n-1} \\ r^n \end{bmatrix} e^{in\theta} \quad ; \quad \begin{bmatrix} [-\frac{n\alpha}{\rho(\omega^2-\omega_T^2)}]r^{-(n+1)} \\ [\frac{in\alpha}{\rho(\omega^2-\omega_T^2)}]r^{-(n+1)} \\ r^{-n} \end{bmatrix} e^{in\theta} \tag{6.9}$$

with $n = 0, 1, 2, \ldots$; $f_n(x)$ is a solution of the Bessel equation of order n [26] and $f'(x)$ denotes the derivative with respect to its argument x. It must be noted that $\omega < \omega_L$ for all the frequencies involved in our problem; thus $Q_L^2 > 0$ and Q_L is always a real quantity (see Eq. (3.37)). On the contrary following (3.33), Q_T is real for $\omega < \omega_T$, but it is an imaginary quantity for $\omega_T < \omega < \omega_L$. An arbitrary solution of equations (3.28) and (3.29) within the conditions of the present problem can be obtained as a linear combination of the vectors given in (6.8) and (6.9). The final results are given by

$$u_r = \frac{e^{in\theta}}{\sqrt{2\pi}} \begin{cases} in/r\, A_1 f_n(Q_T r) + Q_L B_1 f'_n(Q_L r) + \\ \left(n\alpha_a/\rho_a(\omega^2 - \omega_{aT}^2)\right) C_1 r^{n-1} \quad ; \quad r < r_0 \\ \\ in/r\, A_2 f_n(Q_T r) + Q_L B_2 f'_n(Q_L r) - \\ \left(n\alpha_b/\rho_b(\omega^2 - \omega_{bT}^2)\right) C_2 r^{-(n+1)} \quad ; \quad r > r_0 \end{cases} \quad , \quad (6.10)$$

$$u_\theta = \frac{e^{in\theta}}{\sqrt{2\pi}} \begin{cases} -Q_T A_1 f'_n(Q_T r) + in/r\, B_1 f_n(Q_L r) + \\ \left(in\alpha_a/\rho_a(\omega^2 - \omega_{aT}^2)\right) C_1 r^{n-1} \quad ; \quad r < r_0 \\ \\ -Q_T A_2 f'_n(Q_T r) + in/r\, B_2 f_n(Q_L r) + \\ (in\alpha_b/\rho_b(\omega^2 - \omega_{bT}^2)) C_2 r^{-(n+1)} \quad ; \quad r > r_0 \end{cases} \quad , \quad (6.11)$$

and

$$\varphi = \frac{e^{in\theta}}{\sqrt{2\pi}} \begin{cases} (4\pi\alpha_a/\epsilon_{a\infty}) B_1 f_n(Q_L r) + C_1 r^n \quad ; \quad r < r_0 \\ (4\pi\alpha_b/\epsilon_{b\infty}) B_2 f_n(Q_L r) + C_2 r^{-n} \quad ; \quad r > r_0 \end{cases} \quad , \quad (6.12)$$

with $n = 0, 1, 2, \ldots$. The different media are indicated by labels 'a' and 'b' and $f_n(x)$ represents a solution of the Bessel equation of order n which must be bounded in its domain of definition. The constants A_i, B_i, C_i ($i = 1, 2$) are determined through the matching conditions.

While in general the normal modes cannot be obtained as purely longitudinal or transverse, this is possible in the special case $n = 0$ and $q_z = 0$. In this case the condition $\nabla \times \mathbf{u} = 0$ is strictly fulfilled and the modes are purely longitudinal. The solutions $\mathbf{u}$ and φ given by Eqs. (6.10)–(6.12) then coincide which those reported in [10] only for $n = 0$ and $k_z = 0$. In Ref. [10] it is assumed that the modes are always purely longitudinal with $\nabla \times \mathbf{u} = 0$, a fact which we have repeatedly stressed cannot be generally assumed if we are dealing with interfaces. The sole mechanical matching boundary conditions of the continuity of normal velocity and pressure at the interfaces [10], cannot be related to the conditions discussed in Chapter 3 according to the general mathematical treatment of differential equations with piecewise continuous parameters. Nevertheless, if the discussion is limited only to oscillations propagating along the $\mathbf{r}$ direction, then both matching conditions coincide. This is the particular case $n = 0$ and $q_z = 0$, when the longitudinal and transverse solutions can be decoupled.

Confined and interface modes in cylindrical quantum wires

Let us now assume a cylindrical quantum well wire with GaAs for $r < r_0$ and AlAs for $r > r_0$. The complete confinement conditions are (see Chapter 3, and Refs. [27, 28])

$$u_r|_{r=r_0} = 0 \; ; \quad u_\theta|_{r=r_0} = 0 \; ; \tag{6.13}$$

$$\epsilon_{a\infty}\frac{\partial \varphi}{\partial r}|_{r=r_0-} = \epsilon_{b\infty}\frac{\partial \varphi}{\partial r}|_{r=r_0+} \; ; \tag{6.14}$$

$$\varphi|_{r=r_0-} = \varphi|_{r=r_0+} \; , \tag{6.15}$$

and correspond to the mechanically rigid barrier model, in line with the general discussion of Chapters 3 and 4. We stress that only the vibration amplitude $\mathbf{u}$ is confined.

Applying these conditions to Eqs. (6.10)–(6.12) we obtain the completely confined eigenmodes and eigensolutions.

Let us first study the case $n = 0$. Then, we have the transcendental equation

$$J_1(Q_T r_0) J_1(Q_L r_0) = 0 \tag{6.16}$$

with two possible independent solutions which correspond to:

1. Pure longitudinal modes where

$$J_1(Q_L r_0) = 0 \; , \tag{6.17}$$

$J_1(x)$ being the Bessel function of first kind.

2. Pure transverse modes with

$$J_1(Q_T r_0) = 0 \quad . \tag{6.18}$$

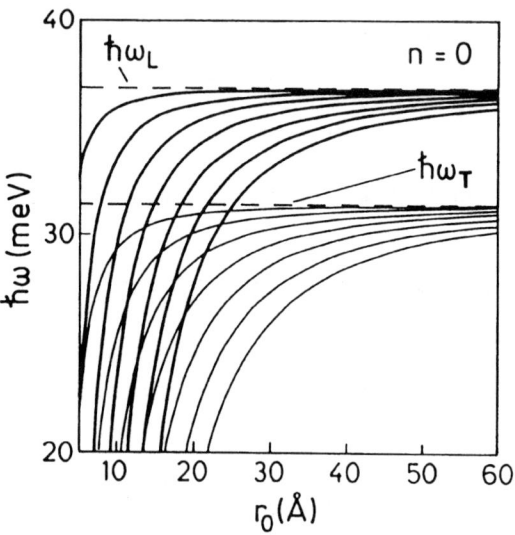

Figure 6.1 The energies of first seven modes in a infinite cylinder with circular cross section as a function of r_0 for the frequency regions below and above ω_T and $n = 0$. The broken lines correspond to bulk longitudinal and transverse phonon energies. The input parameters correspond to a GaAs quantum wire embedded in AlAs.

The corresponding decoupled dispersion relations are

$$\omega^2 = \omega_L^2 - \beta_L^2 \left(\frac{\mu_m}{r_0}\right)^2 \quad ; m = 1, 2 \tag{6.19}$$

and

$$\omega^2 = \omega_T^2 - \beta_T^2 \left(\frac{\mu_m}{r_0}\right)^2 \quad ; m = 1, 2 \ , \tag{6.20}$$

where μ_m ($m = 1, 2, \ldots$) are the zeros of the Bessel function $J_1(x)$ [26]. Figure 6.1 shows the longitudinal — and transverse — mode energies for $n = 0$ as a function of the quantum wire radius r_0. The bulk longitudinal — and transverse-phonon energies are represented by the broken lines for comparison.

The parameters used in the calculations correspond to GaAs according to Table 1.1. The solutions $\mathbf{u}(\mathbf{r})$ for $r < r_0$ and $n = 0$ are given by:

$$\mathbf{u} = B_1 \begin{pmatrix} J_1(\mu_m \frac{r}{r_0}) \\ 0 \\ 0 \end{pmatrix} \quad , \text{if } \omega^2 = \omega_L^2 - \beta_L^2 \left(\frac{\mu_m}{r_0}\right)^2 \qquad (6.21)$$

and

$$\mathbf{u} = B_2 \begin{pmatrix} 0 \\ i J_1(\mu_m \frac{r}{r_0}) \\ 0 \end{pmatrix} \quad , \text{if } \omega^2 = \omega_T^2 - \beta_T^2 \left(\frac{\mu_m}{r_0}\right)^2 , \qquad (6.22)$$

while the associated potential for the pure longitudinal-phonon modes is

$$\varphi = \frac{4\pi \alpha_a B_1}{\epsilon_{a\infty} Q_L} \begin{cases} J_0(\mu_m \frac{r}{r_0}) + \frac{\epsilon_{a\infty}-\epsilon_{b\infty}}{\epsilon_{a\infty}+\epsilon_{b\infty}} J_0(\mu_m) \frac{r}{r_0} \quad ; \quad r < r_0 \\ J_0(\mu_m) + \frac{\epsilon_{a\infty}-\epsilon_{b\infty}}{\epsilon_{a\infty}+\epsilon_{b\infty}} J_0(\mu_m) \frac{r_0}{r} \quad ; \quad r > r_0 \end{cases} \qquad (6.23)$$

and there is of course no electrostatic potential associated with the pure transverse modes (see Figure 6.1).

In Figure 6.2 the vibrational amplitude u_r (6.21) and the electrostatic potential (6.23) are shown for two longitudinal phonon energies $\hbar\omega_1 = 36.5$ meV and $\hbar\omega_2 = 35.9$ meV and $r_0 = 20$ Å.

In the general case with $n \neq 0$ the dispersion relation for the quantum well wire is obtained by solving the following secular equation:

$$\frac{1}{2}(\epsilon_{a\infty} + \epsilon_{b\infty}) \left[J_{n-1}(Q_L r_0) J_{n+1}(Q_T r_0) + J_{n-1}(Q_T r_0) J_{n+1}(Q_L r_0) \right]$$

$$+ \epsilon_{a\infty} \left(\frac{\omega_L^2 - \omega_T^2}{\omega_T^2 - \omega^2}\right) \left[J_{n-1}(Q_L r_0) + \frac{n}{Q_L r_0} J_n(Q_L r_0) \left(\frac{\epsilon_{b\infty}}{\epsilon_{a\infty}} - 1\right) \right] J_{n+1}(Q_T r_0)$$

$$= 0 .$$

(6.24)

The optical frequencies, according to Eq. (6.25), as a function of r_0, are shown in Figures 6.3(a) and 6.3(b) for $n = 1$ and $n = 2$, respectively.

The strong mixing between longitudinal and transverse parts can be clearly seen. For comparison the longitudinal modes in the decoupled model of Ref. [10] (dashed line) is also shown. It can be seen from Figure 6.3 that those curves which resemble the dispersion branches starting with predominantly longitudinal character display an anticrossing with those starting with predominantly transverse character. Then the behaviour changes from quasilongitudinal to

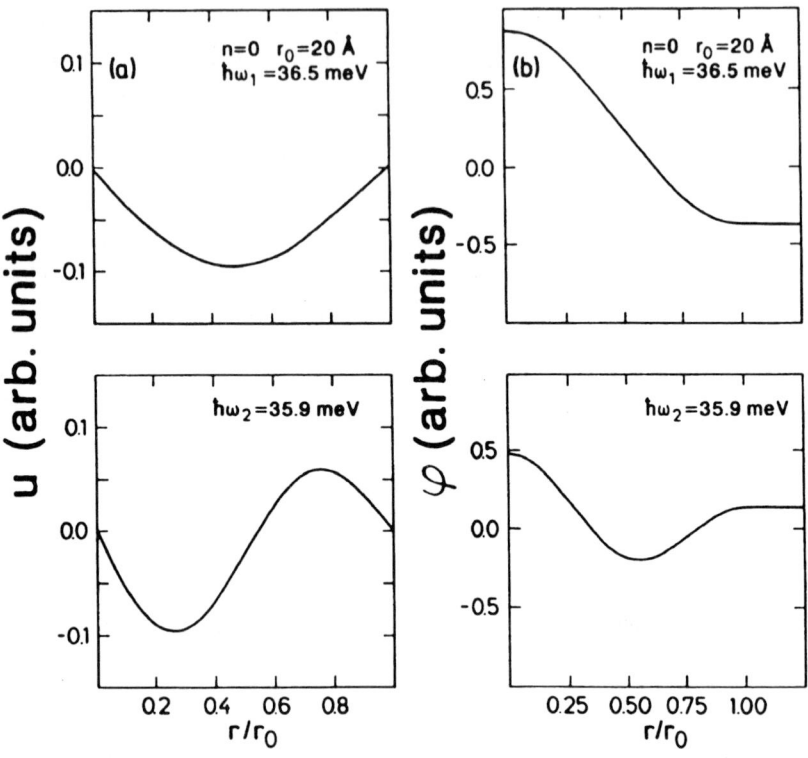

Figure 6.2 The vibrational amplitude u_r (a) and electrostatic potential (b) for $n = 0$ according to Eqs. (6.21) and (6.23), respectively as a function of r/r_0. The calculations were performed for a cylinder of radius $r_0 = 20$ Å and at the phonon energies $\hbar\omega_1 = 36.5$ meV and $\hbar\omega_2 = 35.9$ meV. $\epsilon_{a\infty}$ and $\epsilon_{b\infty}$ were assumed to be equal.

quasitransverse and viceversa. At these points of strong dispersion the contribution from the electric part of the oscillations is stronger. As can be seen in Figure 6.3 these anticrossing are more apparent for small values of r_0. As $r_0 \to \infty$, the bulk longitudinal and transverse phonon dispersion relations are recovered. The study of quantum well wires has been carried out within the simplified assumption $\epsilon_{a\infty} = \epsilon_{b\infty} = \epsilon_\infty$, that is, ignoring the discontinuity in the crystal dielectric background, which is known to have insignificant consequences. In this case the transcendental equation (6.25) is reduced to

$$(\omega_L^2 - \omega^2) J_0(Q_L r_0) J_2(Q_T r_0) + (\omega_T^2 - \omega^2) J_0(Q_T r_0) J_2(Q_L r_0) = 0 \,. \quad (6.25)$$

Quasi-1D semiconductor nanostructures   131

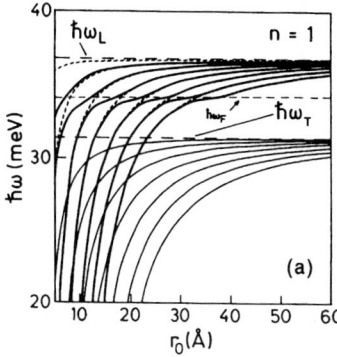
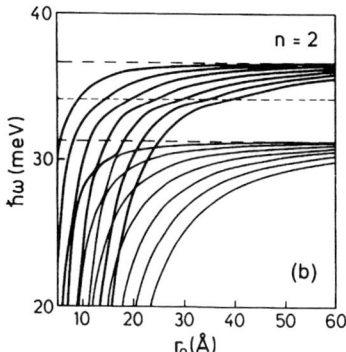

Figure 6.3 The optical vibrational energies of the first seven modes in a GaAs infinite cylinder with circular cross section as a function of r_0 for the frequency region below and above $\omega_0 = \omega_T$ (a) Results for $n = 1$. For comparison, the decoupled longitudinal modes are shown as dashed lines. The horizontal broken lines correspond to the bulk longitudinal and transverse phonon energies. The Fröhlich frequency is denoted as ω_F (b). Results for $n = 2$. For the calculation the parameters corresponding to GaAs were used (see Table 1.1). The approximation $\epsilon_{a\infty} = \epsilon_{b\infty}$ was made.

The eigensolutions $\mathbf{u}_{nm}(r, \theta)$ of the oscillations for the completely confined modes in the quantum well wire are given by:

$$u_r = \frac{B e^{in\theta}}{\sqrt{2\pi}} \left\{ \frac{n}{Q_T r} \frac{J_{n+1}(Q_L r_0)}{J_{n+1}(Q_T r_0)} J_n(Q_T r) + J_n'(Q_L r) + \frac{\epsilon_{a\infty}}{\epsilon_{a\infty}+\epsilon_{b\infty}} \frac{\omega_L^2-\omega_T^2}{\omega_T^2-\omega^2} \right. \quad (6.26)$$

$$\left. \left(J_{n-1}(Q_L r_0) + \frac{n}{Q_L r_0} J_n(Q_L r_0)(\frac{\epsilon_{b\infty}}{\epsilon_{a\infty}} - 1) \right) (\frac{r}{r_0})^{n-1} \right\},$$

$$u_\theta = \frac{iB e^{in\theta}}{\sqrt{2\pi}} \left\{ \frac{J_{n+1}(Q_L r_0)}{J_{n+1}(Q_T r_0)} J_n'(Q_T r) + \frac{n}{Q_L r} J_n(Q_L r) + \frac{\epsilon_{a\infty}}{\epsilon_{a\infty}+\epsilon_{b\infty}} \frac{\omega_L^2-\omega_T^2}{\omega_T^2-\omega^2} \right. \quad (6.27)$$

$$\left. \left(J_{n-1}(Q_L r_0) + \frac{n}{Q_L r_0} J_n(Q_L r_0)(\frac{\epsilon_{b\infty}}{\epsilon_{a\infty}} - 1) \right) (\frac{r}{r_0})^{n-1} \right\},$$

$$\varphi = \frac{4\pi\alpha_a B \, e^{in\theta}}{\epsilon_{a\infty} Q_L \sqrt{2\pi}} \begin{cases} J_n(Q_L r) - \frac{Q_L r_0 \epsilon_{a\infty}}{n(\epsilon_{a\infty}+\epsilon_{b\infty})} \\ \left[J_{n-1}(Q_L r_0) + \frac{n}{Q_L r_0} J_n(Q_L r_0)(\frac{\epsilon_{b\infty}}{\epsilon_{a\infty}} - 1)\right](\frac{r}{r_0})^n \; ; r < r_0 \\ \\ J_n(Q_L r_0) - \frac{Q_L r_0 \epsilon_{a\infty}}{n(\epsilon_{a\infty}+\epsilon_{b\infty})} \\ \left[J_{n-1}(Q_L r_0) + \frac{n}{Q_L r_0} J_n(Q_L r_0)(\frac{\epsilon_{b\infty}}{\epsilon_{a\infty}} - 1)\right](\frac{r_0}{r})^n \; ; r > r_0 \end{cases}$$

(6.28)

and the constant B determines the field amplitude according to the general discussion of Chapter 3.

The amplitudes $\mathbf{u}$ and φ are illustrated in Figure 6.4 for $n = 1$ and $r_0 = 20$ Å and different phonon modes ($\omega < \omega_T$ and $\omega > \omega_T$). As in Figure 6.2, the case of a GaAs quantum well wire embedded in AlAs has been considered.

The first term on the r.h.s. in Eq. (6.26) takes into account the transverse contribution to u_r while the second one in (6.27) represents the longitudinal part in the u_θ component. These mixings are reflected in the optical vibrational energies as a function of r_0 and in the anticrossing points shown in Figures 6.3 (a)-(b) for $n \neq 0$ for $\omega < \omega_T$. These results are in good qualitative agreement with those obtained by microscopic calculation for a quantum well wire with rectangular cross section [20, 21]. The strong mixing between transverse and longitudinal modes has the same character as that described in quantum wells (see Chapter 5).

Figure 6.5 gives the vibrational amplitude $\mathbf{u}$ for $n = 2, r_0 = 21.5$ Å and $\hbar\omega = 36.19$ meV, the first solution of Eq. (6.25). The solid curve corresponds to the $R_{nm}(r)$ component, while the dashed line gives the $\Theta_{nm}(r)$ solution according to Eqs. (6.26), (6.27) and (6.29).

The Fröhlich-like electron-phonon interaction Hamiltonian is then obtained by direct application of the formalism discussed in Chapter 3. In this case we introduced the following labelling and notations of the eigensolutions:

$$\mathbf{u}_{nm}(\mathbf{r}) = (R_{nm}(r)\mathbf{e}_r + \Theta_{nm}(r)\mathbf{e}_\theta) \frac{e^{in\theta}}{\sqrt{2\pi}} . \quad (6.29)$$

The functions $R_{nm}(r)$ and $\Theta_{nm}(r)$ can be easily inferred from the results of (6.26) and (6.27). We define the orthonormality condition so that (see Eq. (3.74)).

$$\int_0^{r_0} \rho \, [R^*_{nm} R_{nm'} + \Theta^*_{nm} \Theta_{nm'}] r \, dr = \delta_{mm'} \quad (6.30)$$

and this determines the constant B of (6.30).

Quasi-1D semiconductor nanostructures 133

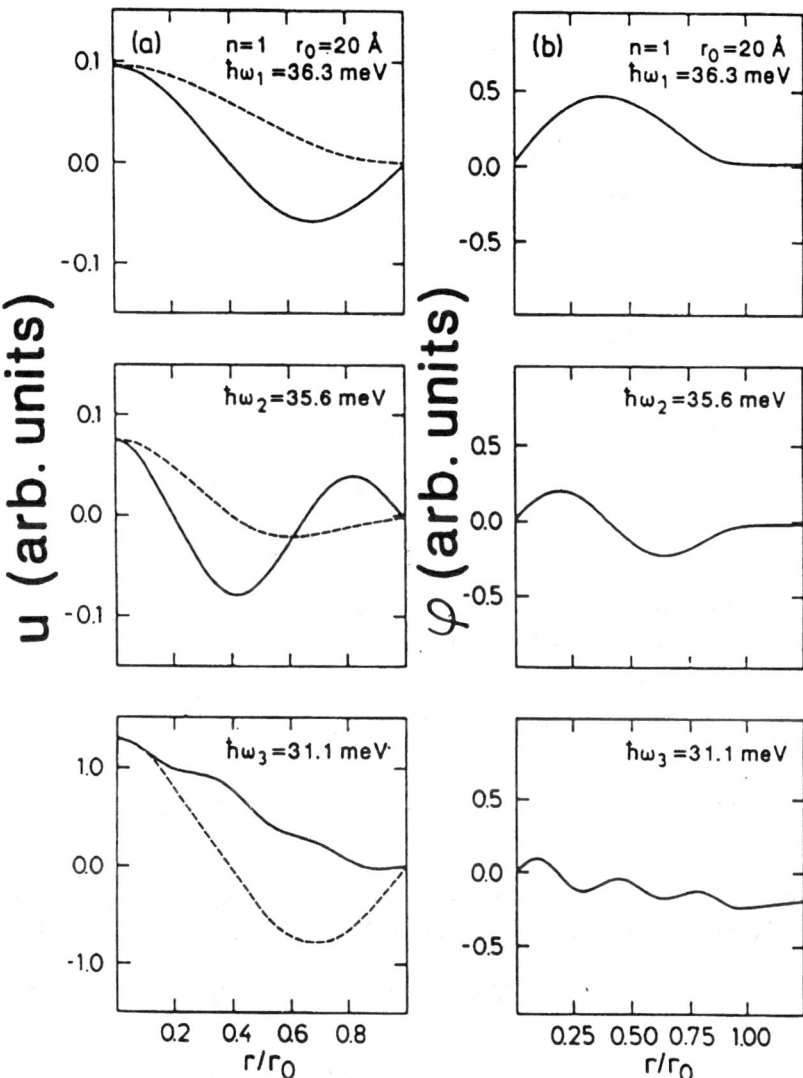

Figure 6.4 The vibrational amplitude displacement **u** (a) and the electrostatic potential φ (b) for $n = 1$ in a GaAs infinite cylinder as a function of r/r_0. The values of $r_0 = 20$ Å and $\hbar\omega_1 = 36.3$ meV, $\hbar\omega_2 = 35.6$ meV ($\omega > \omega_T$), and $\hbar\omega_3 = 31.1$ meV ($\omega < \omega_T$) were used in the calculations. The amplitude R_{1m} is represented by full lines while the Θ_{1m} are given by dashed lines ($m = 1, 2$ and 7).

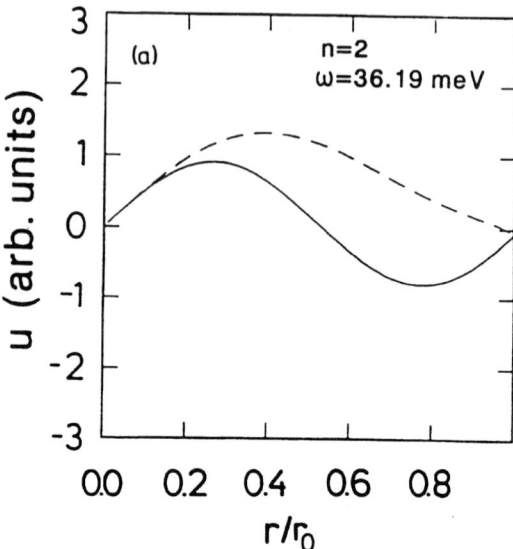

Figure 6.5 The amplitudes $R_{21}(r)$ (solid lines) and $\Theta_{21}(r)$ (dashed line) described by Eqs. (6.26) and (6.27) in a quantum wire as a function of r with $r_0 = 21.5$.

The completeness relation (3.75) is then

$$\sum_{n,m} \rho [R^*_{nm}(r) R_{nm}(r') + \Theta^*_{nm}(r) \Theta_{nm}(r')] e^{in(\theta'-\theta)} = \frac{2\pi}{r} \delta(r-r') \delta(\theta-\theta') .$$

(6.31)

From the obtained eigenvectors the general solution for $\mathbf{u}(r,\theta,t)$ and $\varphi(r,\theta,t)$ in the second quantisation formalism (see Eqs. (3.91) and (3.92)) is given by

$$\hat{u}_r = \sum_{n,m} C_{nm} [R_{nm}(r) e^{i(n\theta - \omega_{nm}t)} \hat{b}_{nm} + H.C.] ,$$

(6.32)

$$\hat{u}_\theta = \sum_{n,m} C_{nm} [\Theta_{nm}(r) e^{i(n\theta - \omega_{nm}t)} \hat{b}_{nm} + H.C.] ,$$

(6.33)

$$\hat{\Phi} = \sum_{n,m} C_{nm} \frac{4\pi \alpha_1}{\epsilon_{a\infty} q_{nm}} [F_{nm}(r) e^{i(n\theta - \omega_{nm}t)} \hat{b}_{nm} + H.C.] ,$$

(6.34)

where the function F_{nm} is determined by the expression for the electrostatic potential (6.29) and $\hat{b}_{nm}$ is the annihilation phonon operator in the state n, m and $k_z = 0$.

The commutation rule (3.97) can then be written as

$$2\rho \sum_{n,m} C_{nm}^2 \omega_{nm} \mathbf{u}_{nm}^*(\rho,\theta) \cdot \mathbf{u}_{nm}(\rho',\theta') = \frac{\hbar}{r}\delta(r-r')\delta(\theta-\theta')\delta(z-z') \ . \quad (6.35)$$

By performing the z-integration in $-\frac{L}{2} \le z \le \frac{L}{2}$, where L is a typical length along the z-axis, the above equation can be cast into the form.

$$2\rho \sum_{n,m} C_{nm}^2 \omega_{nm} \frac{e^{in(\theta-\theta')}}{2\pi} \left[R_{nm}(r)R_{nm}(r') + \Theta_{nm}(r)\Theta_{nm}(r') \right]$$
$$= \frac{\hbar}{r}\delta(r-r')\delta(\theta-\theta') \ . \quad (6.36)$$

If follows from Eqs. (6.31) and (6.36) that

$$C_{nm} = \sqrt{\frac{\hbar}{2\omega_{nm}L}} \ . \quad (6.37)$$

The corresponding Fröhlich-like electron-phonon interaction Hamiltonian $\hat{H} = e\hat{\Phi}$ is given by

$$\hat{H}_F = \sum_{n,m} \bar{C}_F \left(\frac{\pi\rho\omega_L}{\omega_{nm}} \right)^{\frac{1}{2}} \frac{r_0}{q_{nm}} [F_{nm}(r) e^{in\theta} \hat{b}_{nm} + H.C.] \ , \quad (6.38)$$

where $\bar{C}_F$ is $C_F/\sqrt{V}$ and C_F is the Fröhlich constant

$$C_F = \left(e^2 2\pi \hbar \omega_L (\epsilon_\infty^{-1} - \epsilon_0^{-1}) \right)^{\frac{1}{2}}$$

and V is the cylinder volume.

We stress that, within the restrictions of this calculation, the Hamiltonian (6.38) describes the interaction of the electron with the phonons with $k_z = 0$. While this is a particular case, it is in itself important for the study of Raman scattering experiments in quantum wells.

The Hamiltonian (6.38) is reduced to that obtained in the dielectric continuum model for a sufficiently large r_0. Then, in this particular case, the mechanical boundary conditions are irrelevant and the long range interactions associated with the electrostatic potential dominate.

The last term on the r.h.s. of Eqs. (6.26) and (6.27), corresponding to the vector **u**, accounts explicitly for the coupling of electric field with mechanical vibrations. The same can be pointed out for the potential φ given by (6.29). The coupling between φ and **u** by the matching boundary conditions produces the so called interface modes. For large values of r_0 the influence of short range forces on the matching boundary conditions becomes negligible and then the resulting phonon structure is essentially determined by the long range electrostatic forces. As $r_0 \to \infty$ the bulk transverse and longitudinal phonon dispersion relations are recovered while for finite r_0 a new solution appears between $\hbar\omega_L$ and $\hbar\omega_T$ - the so-called Fröhlich frequency ω_F. This corresponds to a homogeneous polarisation of the cylinder and results from the electrostatic matching boundary conditions in finite media [29]. For that frequency, there is no difference between longitudinal and transverse modes. For the cylindrical geometry ω_F takes the form

$$\omega_F = \sqrt{\frac{\epsilon_{a0} + \epsilon_{b\infty}}{\epsilon_{a\infty} + \epsilon_{b\infty}}}\,\omega_T . \tag{6.39}$$

In Figure 6.3a ω_F is shown for $\epsilon_{a\infty} = \epsilon_{b\infty}$ with the GaAs parameters $\hbar\omega_F = 34.15$ meV. Hence in this particular case, under these circumstances, the phenomenological model here presented can be reduced to the dielectric continuum one with electrodynamical matching boundary conditions as a limiting case but if r_0 is smaller than or of the order of Q_L^{-1} and Q_T^{-1}, then the effects of the mechanical boundary conditions become important and one must observe full matching boundary conditions.

Free-standing wires

For a free standing wire the normal mode oscillations can be considered as free at the surface S and the condition $\boldsymbol{\sigma} \cdot \mathbf{N} = 0$ at $\mathbf{r} \in S$ is required. Then the vector **u** is only defined in the active medium and we have $\mathbf{u} \equiv 0$ for $r > r_0$ but there is an electrostatic potential $\varphi \neq 0$ in the entire space.

The vector $\boldsymbol{\sigma}_N = \boldsymbol{\sigma} \cdot \mathbf{e}_r$, defined in (3.27), in cylindrical coordinates has the form

$$\boldsymbol{\sigma}_N = \sigma_r \mathbf{e}_r + \sigma_\theta \mathbf{e}_\theta + \sigma_z \mathbf{e}_z , \tag{6.40}$$

where

$$\sigma_r = -\rho\beta_L^2 \frac{\partial u_r}{\partial r} - \rho(\beta_L^2 - 2\beta_T^2)(\frac{\partial u_z}{\partial z} + \frac{1}{r}\frac{\partial u_\theta}{\partial \theta} + \frac{1}{r}u_r),$$

$$\sigma_\theta = -\rho\beta_T^2(\frac{\partial u_\theta}{\partial r} + \frac{1}{r}\frac{\partial u_r}{\partial \theta} - \frac{1}{r}u_\theta), \tag{6.41}$$

$$\sigma_z = -\rho\beta_T^2(\frac{\partial u_r}{\partial z} + \frac{\partial u_z}{\partial r}).$$

The matching boundary conditions corresponding to the present case are:

- continuity of φ at $r = r_0$

- $\sigma_N = 0$ at $r = r_0$

- continuity of $D_N \equiv D_r$ at $r = r_0$

The last condition entails the following relation

$$4\pi \alpha_a u_r(r_0-, \theta) - \epsilon_{a\infty}\frac{\partial \varphi}{\partial r}(r_0-, \theta) = -\epsilon_{b\infty}\frac{\partial \varphi}{\partial r}(r_0+, \theta) \quad . \quad (6.42)$$

The matching conditions described above should be applied to the general solutions given by Eqs. (6.10)–(6.12). This leads to the following solution

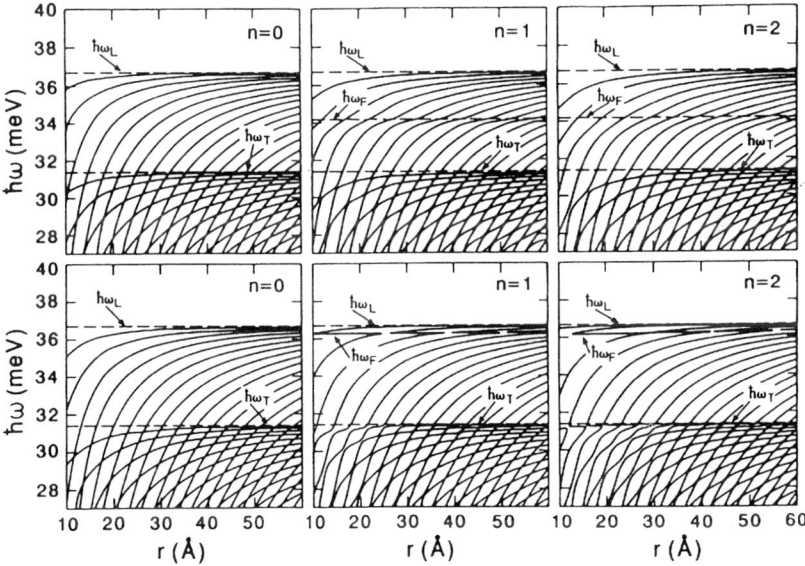

Figure 6.6 Optical vibrational energies $\hbar\omega$ as a function of the wire radius r_0 in a free standing wire. For comparison the completely confined quantum well wire case is shown in the upper part of the figure for the same values of n ($n = 0, 1$ and 2). The dashed lines correspond to the longitudinal and transverse bulk frequencies. The Fröhlich frequency ω_F is indicated by dashed lines. For the calculation, the parameters corresponding to GaAs have been used.

$$u_r = \frac{B\,e^{in\theta}}{\sqrt{2\pi}} \left[-\frac{n}{Q_T r_0^2 r} \frac{g_n(Q_L)}{J_{n+2}(Q_T r_0)} J_n(Q_T r) + Q_L J'(Q_L r) \right.$$
$$\left. + n r_0 \left(\frac{\beta_L^2}{\beta_T^2} Q_L^2 - Q_T^2 \right) t_n(Q_L, Q_T) \left(\frac{r}{r_0} \right)^{n-1} \right] \quad , \quad (6.43)$$

$$u_\theta = i \frac{B\,e^{in\theta}}{\sqrt{2\pi}} \left[-\frac{1}{Q_T r_0^2} \frac{g_n(Q_L)}{J_{n+2}(Q_T r_0)} J'_n(Q_T r) + \frac{n}{r} J_n(Q_L r) \right.$$
$$\left. + n r_0 \left(\frac{\beta_L^2}{\beta_T^2} Q_L^2 - Q_T^2 \right) t_n(Q_L, Q_T) \left(\frac{r}{r_0} \right)^{n-1} \right] \quad , \quad (6.44)$$

$$\varphi = \frac{4\pi\alpha}{\epsilon_{a\infty}} \frac{B\,e^{in\theta}}{\sqrt{2\pi}} \begin{cases} J_n(Q_L r) - t_n(Q_L, Q_T) \left(\frac{r}{r_0} \right)^n & ; r < r_0 \\ J_n(Q_L r_0) - t_n(Q_L, Q_T) \left(\frac{r_0}{r} \right)^n & ; r > r_0 \end{cases} , \quad (6.45)$$

where

$$g_n(Q_L) = \left(\frac{\beta_L}{\beta_T} \right)^2 (Q_L r_0)^2 J_n(Q_L r_0) - 2(n+1) Q_L r_0 J_{n+1}(Q_L r_0) \quad , \quad (6.46)$$

and

$$t_n(Q_L, Q_T) = \left[(Q_T r_0)^2 J_n(Q_L r_0) J_{n+2}(Q_T r_0) + \frac{\epsilon_{a\infty}}{\epsilon_{b\infty}} g_n(Q_L) J_n(Q_T r_0) \right]$$
$$\left[\left(\frac{\epsilon_{a\infty}}{\epsilon_{b\infty}} \frac{\beta_L}{\beta_T} \right)^2 (Q_L r_0)^2 + (Q_T r_0)^2 \right) J_{n+2}(Q_T r_0) \right]^{-1} . \quad (6.47)$$

with a normalisation constant B as in the previous case. The secular equation yielding the eigenfrequencies is:

$$2n(n-1)(\frac{r_0}{\beta_T})^2 (\omega_L^2 - \omega_T^2) t_n(Q_L, Q_T) J_{n+2}(Q_T r_0) + 2n(Q_T r_0)^2$$
$$J_{n+2}(Q_T r_0) \left[Q_L r_0 J'_n(Q_L r_0) - J_n(Q_L r_0) \right] + g_n(Q_L) \left[2 Q_T r_0 J'_n(Q_T r_0) \right.$$
$$\left. + (Q_T^2 r_0^2 - 2n^2) J_n(Q_T r_0) \right] = 0 . \quad (6.48)$$

Figure 6.6 shows the phonon energies $\hbar\omega$ as a function of radius r_0 for three values of n: $n = 0, 1, 2$. For comparison the quantum well wire case is displayed in the upper part of the figure. The three graphs of the lower part correspond to the free standing wire. For $n = 0$ we have decoupled longitudinal and transverse modes. For $n \neq 0$ the modes are coupled and as in the completely confined case, show a strong mixing between the transverse and longitudinal parts of the oscillations. Anticrossings take place for $n = 1, 2$ where the curves change their behaviour from quasilongitudinal to quasitransverse optical modes and viceversa. For the numerical calculation the parameters for GaAs of Table 1.1 were used with $\epsilon_{b\infty} = 1$. The Fröhlich frequency ω_F is indicated by dashed lines and substituting the GaAs parameters in Eq. (6.39) yields $\hbar\omega_F = 36.28$ meV.

Figure 6.7 shows the components of the vibrational amplitude vector $\mathbf{u}$ as a function of r for $n = 0, 1, 2$ and a radius of $r_0 = 21.5$ Å. The first and second solutions of Eq. (6.48) for given n: ($n = 0$, 1 and 2) are shown. The solid curve corresponds to the $R_{nm}(r)$ component ($u_r = B e^{in\theta} R_{nm}(r)/\sqrt{2\pi}$), while the dashed line to the Θ_{nm} component ($u_\theta = i B e^{in\theta} \Theta_{nm}(r)/\sqrt{2\pi}$). The function $\Theta_{nm}(r)$ vanishes in the $n = 0$ case if the frequency corresponds to longitudinal solutions. The amplitude at $r = r_0$ can be different from zero because the free surface conditions are applied to the stress component, i.e. $\sigma_N = 0$ at $r = r_0$.

Following the procedure described in Chapter 3 and section 6, the electrostatic potential and Fröhlich electron-phonon Hamiltonian are given in this case by

$$\hat{H}_F = e \sum_{n,m} \left[\Phi_{n,m}(r,\theta) \hat{b}_{nm} + H.C. \right]$$

$$= \sum_{n,m} C_{nm} \left[F_{nm}(r) e^{in\theta} \hat{b}_{nm} + H.C. \right] , \quad (6.49)$$

where

$$C_{nm} = \left[\frac{\pi \omega_L \rho}{\omega_{nm}} \right]^{\frac{1}{2}} r_0^2 \bar{C}_F \quad (6.50)$$

and

$$F_{nm}(r) = B_{nm} \begin{cases} J_n(Q_L r) - t_n(Q_L, Q_T)(\frac{r}{r_0})^n ; & r < r_0 \\ vv \left[J_n(Q_L r_0) - t_n(Q_L, Q_T) \right] (\frac{r_0}{r})^n ; & r > r_0 \end{cases} , \quad (6.51)$$

where $t_n(Q_L, Q_T)$ was defined in (6.47) and B_{nm} is the normalisation constant to ensure the orthonormality of $\mathbf{u}_{nm} = (u_r, u_\theta)$.

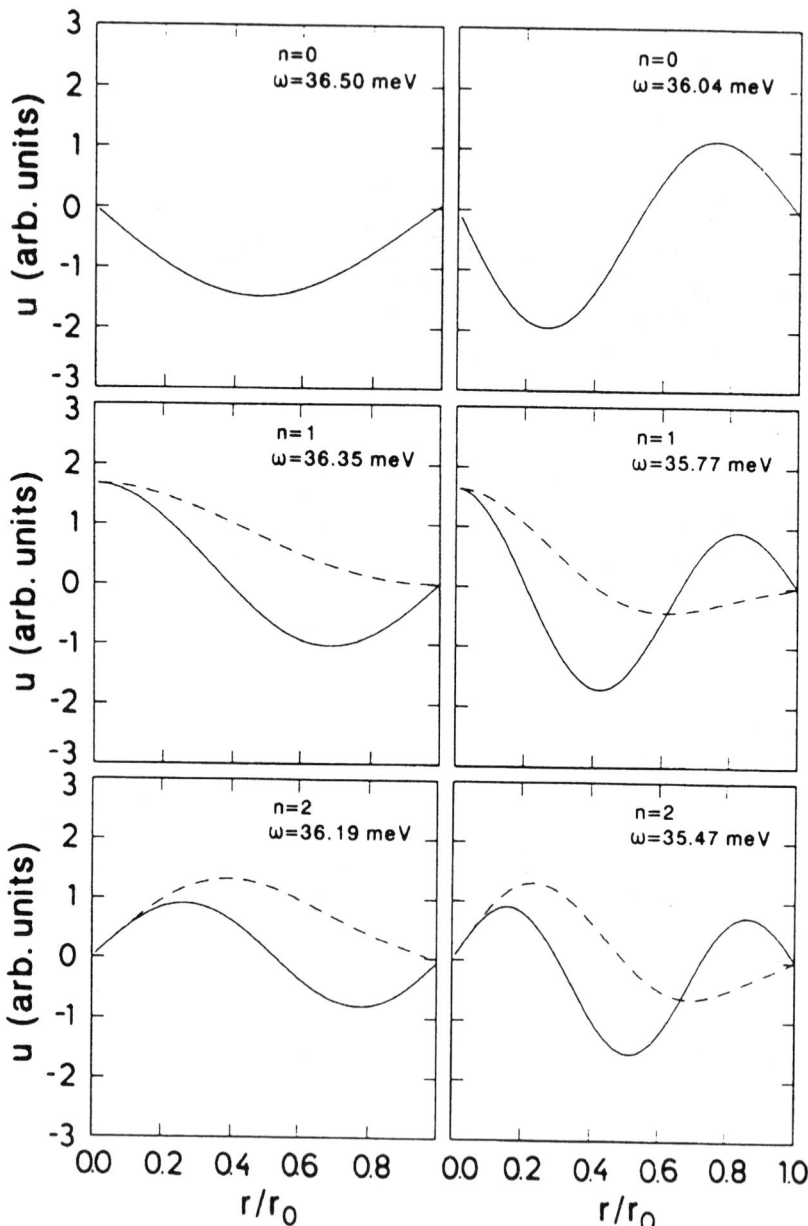

Figure 6.7 The components of the vibration amplitude vector **u** as a function of r for $n = 0$, 1 and 2; $m = 1$ (l.h.s.) and 2 (r.h.s.) for the free standing wire case. Vibrational amplitudes $R_{rm}(r)$ (solid line) and $\Theta_{nm}(r)$ (dashed lines). The calculations correspond to a $r_0 = 21.5$ Å. The amplitudes have been normalised following Eq. (6.30). The same arbitrary units are used in all figures and $\epsilon_{b\infty} = 1$ everywhere.

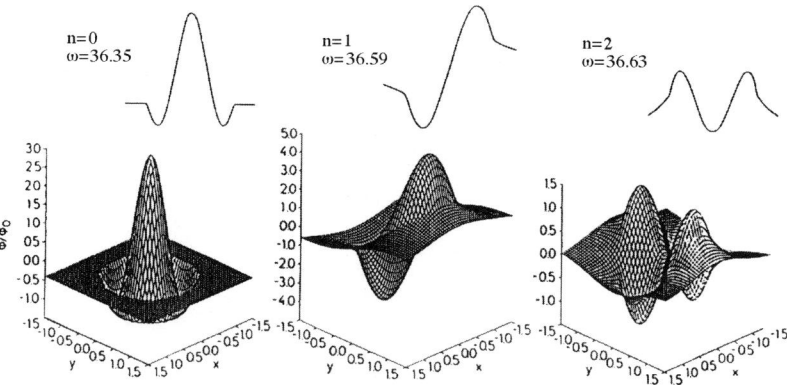

Figure 6.8 Three-dimensional plots of the electrostatic potential $\Phi_{nm}(r,\theta)$ as a function of $x = r/r_0 \cos\theta$ and $y = r/r_0 \sin\theta$ for $n = 0$, 1 and 2 ($m = 1$). The electrostatic potential has been divided by the constant $(2\pi\hbar^2\omega_L^2(\epsilon_\infty^{-1} - \epsilon_0^{-1})/V)^{1/2}$, and has units of meV^{-1} Å. The parameters for GaAs are those of Table 1.1 and $r_0 = 21.5$ Å. The dielectric constant $\epsilon_{b\infty}$ of the surrounding medium is taken equal to unity. The profiles in the $x = 0$ plane are shown in the upper part of the figure.

In Figure 6.8 we show the 3D plots for the electrostatic potential $\varphi_{nm}(r,\theta)$. We also present the $n = 0$, 1 and 2 modes for $m = 1$ ($\hbar\omega_{01} = 36.35$ meV, $\hbar\omega_{11} = 36.59$ meV and $\hbar\omega_{21} = 36.63$ meV). In the plots we use variables $x = r\cos\theta$, $y = r\sin\theta$, measured in units of r_0. The potential profiles in the upper part of Figure 6.8 were taken at $x = 0$, while φ is measured in units of $\Phi_0 = -C_F/e(\hbar\omega_L)^{\frac{1}{2}}$. We can observe the continuity of the potential at $r = r_0$. In Figure 6.9 the equipotential curves in the (x, y) plane (top) and as a three-dimensional plot (bottom) are shown. The case $n = 2$ and $m = 2$ is shown for $r_0 = 21.5$ Å.

In the two cases here analysed (completely confined and free standing wire) these results have the correct symmetry pattern for the electrostatic potential and the vibrational amplitudes while the mechanical and electrostatic conditions at the surface/interface are consistently observed.

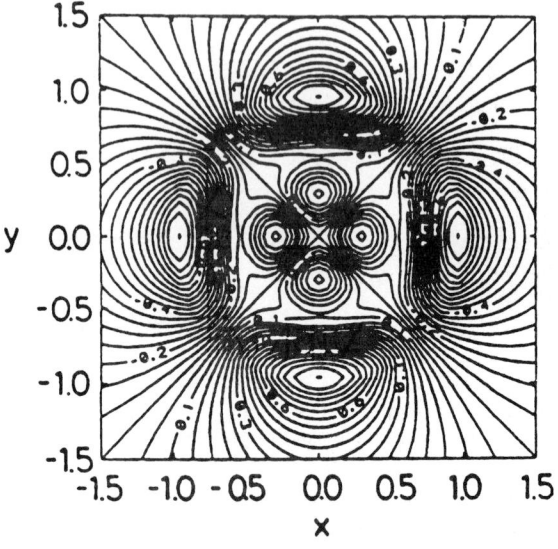

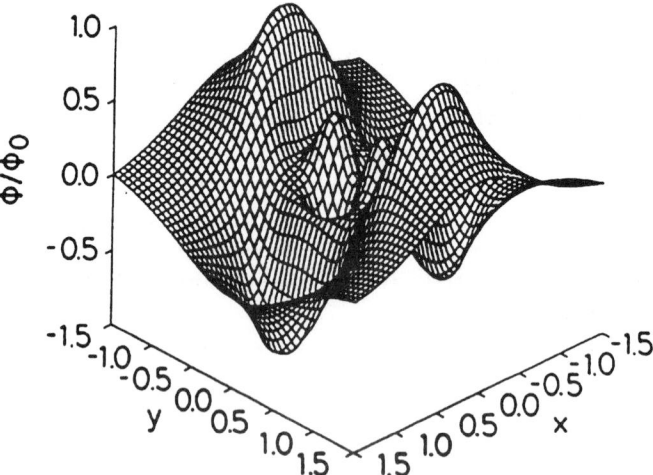

Figure 6.9 Electrostatic potential $\Phi_{nm}(r,\theta)$ as a function of $x = r/r_0 \cos\theta$ and $y = r/r_0 \sin\theta$ for $n = m = 2$. At the top we show the projection on the $z = 0$ plane equipotential curves, while the bottom gives the 3D plot. The potential has been divided by the constant $\Phi_0 = -C_F/e(\hbar\omega_L)^{\frac{1}{2}}$ and Φ_{nm}/Φ_0 has units of meV$^{\frac{1}{2}}$ Å. The same parameters as in Figure 6.8 were used in the calculation.

References

1. M. Watt, C.M. Sotomayor-Torres, H.E.G. Arnot, and S.P. Beaumont, Semicond. Sci. Technol. **5**, 285 (1990).

2. P.M. Petroff, A.C. Gossard, R.A. Logan, and W. Wiegmann, Appl. Phys. Lett. **41**, 635 (1982).

3. T.J. Thornton, M. Pepper, H. Ahmed, D. Andrews, and G.J. Davies, Phys. Rev. Lett. **56**, 1198 (1986).

4. K. Ismail, D.A. Antoniadis, and H.I. Smith, Appl. Phys. Lett. **54**, 1130 (1989).

5. M. Tsuchiya, J.M. Gaines, R.H. Yan, R.J. Simes, P.O. Holtz, and P.M. Petroff, Phys. Rev. Lett. **62**, 466 (1989).

6. D.G. Hasko, A. Potts, J.R.A. Cleaver, C.G. Smith, and H. Ahmed, J. Vac. Sci. Technol. **B6**, 1849 (1989).

7. *Nanostructure Physics and Fabrication*, edited by M.A. Reed and P. Kirk, Academic, Boston (1989).

8. C.D.W. Wilkinson and S.P. Beaumont in: *The Physics and Fabrication of Microstructures and Microdevices*, (Springer Proccedings in Physics) V.13 ed. C. Weisbuch and M.J. Kelly (Springer, Berlin), p.36.

9. A.M. Stroscio, Phys. Rev. **B40**, 6428 (1989).

10. N.C. Constatinou and B.K. Ridley, Phys. Rev. **B41**, 10622 (1990).

11. A.M. Stroscio, V.K. Kim, A.M. Littlejohn, and H. Chuang, Phys. Rev. **B42**, 1488 (1990).

12. V.B. Campos, S. Das Sarma and M.A. Stroscio, Phys. Rev. **B46**, 3849 (1992).

13. R. Enderlein, Phys. Rev.**B47**, 2162 (1993).

14. P. Selbmann and R. Enderlein, Superlattices and Microstructures **12**, 219 (1992).

15. P.A. Knipp and T.L. Reinecke, Phys. Rev. **B45**, 9091 (1992).

16. P.A. Knipp and T.L. Reinecke, Phys. Rev. **B48**, 5700 (1993).

17. X.F. Wang and X.L. Lei, Phys. Rev. **B49**, 4780 (1994).

18. N. Constatinou and B.K. Ridley, Phys. Rev. **B41**, 10627 (1990).

19. K.W. Kim, M.A. Stroscio, A. Bhatt, R. Mickevicius, and V.V. Mitin, J. Appl. Phys. **70**, 319 (1991).

20. S.F. Ren and Y.C. Chang, Phys. Rev. **B43**, 11857 (1991).

21. F. Rossi, L. Rota, C. Bungaro, P. Lugli and E. Molinari, Phys. Rev. **B47**, 1695 (1993).

22. B. Zhu, Phys. Rev. **B44**, 1926 (1991).

23. B. Zhu, Sci. Technol. **B7**, 88 (1992).

24. G. Fasol, M. Tanaka, H. Sakuki and Y. Horikoshi, Phys. Rev. **B38**, 6056 (1988).

25. B. Jusserand and M. Cardona in *Light Scattering in Solids V* (Topics in Applied Physics 66) ed. M. Cardona and G. Güntherodt (Heidelberg, Springer), p.61.

26. M. Abramovitz and I.A. Stegun, *Handbook of Mathematical Functions*, Dover, New York (1972).

27. F. Comas, C. Trallero-Giner, and A. Cantarero in 'Phonon in Semiconductor Nanostructures', NATO ASI Series E, V. 236. ed. J.P. Leburton, J. Pascual and C. Sotomayor-Torres, Kluwer Academic Publs., Dordrecht (1993), p.49.

28. F. Comas, C. Trallero-Giner and A. Cantarero, Phys. Rev. **B47**, 7602 (1993).

29. R. Ruppin and R. Engelman, Rep. Prog. Phys. **33**, 149 (1970).

CHAPTER SEVEN

Quasi-OD semiconductor nanostructures

Confined systems with spherical geometry — quantum dots — are here discussed with an approach similar to that used in Chapter 6 to study quantum wires.

Besides the quantum wires, discussed in Chapter 6, contemporary advanced technologies permit the growth of semiconductor nanostructures with quasi-zero-dimensional properties, also known as quantum dots. The diameter of these structures is in the range of a few nanometers. The possibilities of realising new devices in these novel systems has stimulated the investigation of the vibrational properties of III-V semiconductor materials a well as microcrystallites of II-VI embedded in glass [1]. Characteristic physical properties appear which suggest a broad spectrum of device applications for electro-optical and nonlinear optical applications in communications and computing, such as long wavelength optical filters and switching devices [2]. Quantum dots based on $Si/Si_{0.91}Ge_{0.09}$ and Si_9Ge_9 strained superlattices have been fabricated using electron beam lithography and reactive ion etching [3]. Solid phase precipitation of II-VI and I-VII semiconductors from a solid or liquid solution is one of the widely used methods to grow microcrystallites (See Reference [4] and references therein). Quantum dots of materials such as CdS [5, 6], CdSe [7–11], CdSSe [12–14], $(CdS)_x(CdSe)_{1-x}$ [15], CdZnSe [4], PbS [16, 17], GaAs [18], GaP [19, 20], InSb [21–23], GaSb [22, 23], AlSb [23], InAs, InP, InGaAs [22], Si [24–26], Ge [27], PbI_2 [28], SiC [29], etc. have been obtained and their optical and vibrational properties studied.

Fröhlich was the first to study the vibrational modes in spherical microscrystallites and derived the so-called Fröhlich frequency [30]. Macroscopic treatments of optical modes in small spheres have been described [31, 32] based on a dielectric model with no account of mechanical boundary conditions. Models based on phenomenological continuum approaches have been proposed for the study of polar optical oscillation in semiconductor quantum dots [8, 9, 11, 15, 27]. Longitudinal vibrational modes and surface optical modes have been obtained by using the classical dielectric model or the so called hydrodynamic model [33] upon application of spherical boundary conditions. These treatments are all subject to the limitations discussed in Chapter 3 and 4 for heterostructures in general. In this chapter, we discuss the application of the long wave macroscopic model to the study of the normal modes in spherical quantum dots or microcrystallites, following the general line of argument of Chapter 3. We take full account of both spatial and electrostatic dispersion (within the model) as well as of mechanical and electrostatic matching boundary conditions. We shall study the optical mode dispersion and general analytical expressions for the electrostatic potential and mechanical vibration amplitude for a spherical quantum dot. The Fröhlich optical interaction Hamiltonian will also be given. We illustrate the optical mode dispersion relations for CdS microcrystallites in glass and a GaAs quantum dot in an AlAs matrix.

Linearly independent solutions for a spherical quantum dot

Consider a single spherical quantum dot of radius R. We use spherical coordinates (r, θ, φ) centred in the centre of the sphere. To obtain the general solution it is necessary to solve again in the new geometry the Helmholtz equations for Ψ and Γ and the Laplace equation for φ_H within each medium and to match them at the interfaces. For a spherical quantum dot the bounded solution of Ψ and φ_M described by Eqs. (3.36) and (3.39), respectively are given by

$$\varphi_H = Y_{lm}(\theta, \varphi) \begin{cases} C_1 r^l & ; \quad r < R \\ C_2 r^{-l-1} & ; \quad r > R \end{cases} \tag{7.1}$$

and

$$\Psi = Y_{lm}(\theta, \varphi) \begin{cases} A_1 j_l(Q_L r) & ; \quad r < R \\ A_2 h_l^{(1)}(Q_L r) & ; \quad r > R \end{cases}, \tag{7.2}$$

where $Y_{lm}(\theta, \varphi)$ with $m = -l, \ldots l$ are the spherical harmonics, $h_l^{(1)}$ the Hankel spherical functions and C_i and A_i ($i = 1, 2$) constants to be determined through the standard matching analysis and, furthermore, by determining the field amplitude according to the normalisation procedure described in Chapter 3. As usual, for the definitions of the spherical functions we follows the notation of Ref. [34].

By combining Eqs. (7.1), (7.2) and (3.38) the scalar potential can be written as

$$\varphi(r, \theta, \varphi) = Y_{lm}(\theta, \varphi) \begin{cases} -\dfrac{4\pi \alpha_a}{\epsilon_{a\infty} Q_{La}^2} A_1 j_l(Q_{La} r) + C_1 r^l & r < R \\ -\dfrac{4\pi \alpha_b}{\epsilon_{b\infty} Q_{Lb}^2} A_2 h_l^{(1)}(Q_{Lb} r) + C_2 r^{-l-1} & r > R \end{cases}.$$

$$\tag{7.3}$$

The parameters α_i, Q_{Ti} and $\epsilon_{i\infty}$ ($i = a, b$) take values corresponding to each bulk constituent medium ($i = a(b)$ inside (outside) the sphere) although, as repeatedly pointed out, the effects of the discontinuities in the crystal background dielectric constant are unimportant and will in practice be neglected.

The solution of the vector equation (3.32) for the auxiliary vector $\mathbf{\Gamma}$ can be obtained by introducing the natural coordinates for the treatment of a spherical quantum dot. The vectorial character of Eq. (3.32) leads to two kinds of complications: firstly the structure of $\nabla^2 \mathbf{\Gamma}$ is complicated in curvilinear coordinates; secondly, owing to the fact that $\mathbf{\Gamma} = \nabla \times \mathbf{u}$, the components of $\mathbf{\Gamma}$ are not independent but are related by the equation

$$\nabla \cdot \mathbf{\Gamma} = 0 . \tag{7.4}$$

If $\mathbf{e}_i$ ($i = 1, 2, 3$) is a curvilinear coordinate basis, then only two components of $\mathbf{\Gamma}$ are independent because of Eq. (7.4). Thus, it suffices to obtain general solutions for two of these components $\Gamma_i \mathbf{e}_i$, of $\mathbf{\Gamma}$. That is to say, the determination of $\mathbf{\Gamma}$, which must be a solution of (3.32) requires simply the determination of two scalar functions. A general procedure to do this is the following: Consider two vectors $\mathbf{M}$ and $\mathbf{N}$ of the form:

$$\mathbf{M} = \nabla \times (v_1 \mathbf{r}) \tag{7.5}$$

and

$$\mathbf{N} = \frac{1}{Q_T} \nabla \times \nabla \times (v_2 \mathbf{r}) . \tag{7.6}$$

These are independent solutions of Eqs. (3.32) and (7.4) if v_1 and v_2 satisfy the scalar Helmholtz equation [35]

$$(\nabla^2 + Q_T^2) v_i = 0 , \quad (i = 1, 2) . \tag{7.7}$$

Having determined the two vectors $\mathbf{M}$ and $\mathbf{N}$, we then obtain the vector function $\mathbf{\Gamma}$ as

$$\mathbf{\Gamma} = \mathbf{M} + \mathbf{N} . \tag{7.8}$$

The bounded solutions of (7.7) for $r < R$ can be written as

$$v = C j_l(Q_T r) Y_{lm}(\theta, \varphi) \tag{7.9}$$

and C is a constant.

The operator ∇ has the form [36]

$$\nabla = \mathbf{e}_r \frac{\partial}{\partial r} - \frac{i}{r} \mathbf{e}_r \times \hat{\ell}, \tag{7.10}$$

where $\hat{\ell} = -i\mathbf{r} \times \nabla$ and $\mathbf{e}_r$ is a unitary vector along the $\mathbf{r}$ direction. We introduce the following notation:

$$\mathbf{X}_{lm} = \frac{\hat{\ell} Y_{lm}}{\sqrt{l(l+1)}}, \quad l \neq 0. \tag{7.11}$$

In a spherical basis $\mathbf{X}_{lm}$ can be cast as

$$\mathbf{X}_{lm}(\theta,\varphi) = \frac{\sqrt{l(l+1)}}{(2l+1)\sin\theta} \times \left[m\frac{(2l+1)}{l(l+1)} i Y_{lm} \mathbf{e}_\theta - \left(\frac{l-m+1}{l+1} Y_{(l+1)m} - \frac{l+m}{l} Y_{(l-1)m}\right) \mathbf{e}_\varphi \right]. \tag{7.12}$$

For $l = 0$, because Y_{00} is a constant, we define $\mathbf{X}_{00} = 0$. The vectors $\mathbf{M}$ and $\mathbf{N}$ satisfy the vector identities:

$$\mathbf{M} = (\nabla v_1) \times \mathbf{r}, \tag{7.13}$$
$$\nabla \times \mathbf{N} = Q_T (\nabla v_2) \times \mathbf{r}, \tag{7.14}$$
$$\nabla \times \nabla \times \mathbf{N} = \nabla\nabla \cdot \mathbf{N} + Q_T^2 \mathbf{N}. \tag{7.15}$$

In order to calculate the vector $\mathbf{u}$ it is necessary to obtain $\mathbf{M}$ and $\nabla \times \mathbf{N}$. These vectors can be found from Eqs. (7.13) and (7.14) by making use of Eq. (7.10):

$$\mathbf{M} = -i B_1 \sqrt{l(l+1)} j_l(Q_T r) \mathbf{X}_{lm}, \tag{7.16}$$

$$\nabla \times \mathbf{N} = -i Q_T D_1 \sqrt{l(l+1)} j_l(Q_T r) \mathbf{X}_{lm}, \tag{7.17}$$

where B_1 and D_1 are constants.

The vector $\mathbf{\Gamma}$ can be obtained by combining the vector identity (7.15) and the equality (7.8). Thus

$$\mathbf{\Gamma} = \mathbf{M} + \frac{1}{Q_T^2}(\nabla \times \nabla \times \mathbf{N} - \nabla\nabla \cdot \mathbf{N}) \tag{7.18}$$

and the final expression is:

$$\mathbf{\Gamma} = -i B_1 \sqrt{l(l+1)} j_l(Q_T r) \mathbf{X}_{lm} - i D_1 \frac{\sqrt{l(l+1)}}{Q_T r}$$
$$\left[i\sqrt{l(l+1)} j_l(Q_T r) Y_{lm} \mathbf{e}_r + \frac{d}{dr}(r j_l(Q_T r)) \mathbf{e}_r \times \mathbf{X}_{lm} \right]. \tag{7.19}$$

To obtain $\nabla \times \mathbf{\Gamma}$ we note that the evaluation of $\nabla \times \mathbf{M}$ can be simply performed by using (7.10) in (7.16) and according to (7.11) we have

$$\nabla \times \mathbf{\Gamma} = -iB_1 \frac{\sqrt{l(l+1)}}{r} \left[i\sqrt{l(l+1)} j_l(Q_T r) Y_{lm} \mathbf{e}_r \right.$$
$$\left. + \frac{d}{dr}(r j_l(Q_T r)) \mathbf{e}_r \times \mathbf{X}_{lm} \right] - iQ_T D_1 \sqrt{l(l+1)} j_l(Q_T r) \mathbf{X}_{lm} . \quad (7.20)$$

Hence, the curvilinear coordinate basis to describe the oscillation modes in a spherical quantum dot appears naturally and is defined by $(\mathbf{e}_r, \mathbf{X}_{lm}, \mathbf{e}_r \times \mathbf{X}_{lm})$. The vector $\mathbf{u}$ can then be written in the form.

$$\mathbf{u} = u_1 Y_{lm} \mathbf{e}_r + u_2 \mathbf{X}_{lm} + u_3 \mathbf{e}_r \times \mathbf{X}_{lm} . \quad (7.21)$$

Substituting Eqs. (7.1), (7.3), and (7.20) in Eq. (3.32) yields general expression for the u_i ($i = 1, 2$ and 3). For $r \leq R$ these components are given by:

$$u_1 = \left[-\frac{A_1}{Q_L^2} \frac{d}{dr} (j_l(Q_L r)) + \frac{B_1 l(l+1)}{r Q_T^2} j_l(Q_T r) - \frac{\alpha C_1 l}{\rho \beta_T^2 Q_T^2} l r^{l-1} \right] Y_{lm} ,$$

$$u_2 = -\frac{D_1}{Q_T} \sqrt{l(l+1)} j_l(Q_T r) ,$$

$$u_3 = -i \frac{\sqrt{l(l+1)}}{r} \left[-\frac{\alpha C_1}{\rho \beta_T^2 Q_T^2} r^l - \frac{A_1}{Q_L^2} j_l(Q_L r) + \frac{B_1}{Q_T^2} \frac{d}{dr} (r j_l(Q_T r)) \right] .$$

$$(7.22)$$

In the above formulae the function $j_l(Q_T r)$ is transformed into $i_l(Q_T r)$ for $Q_T^2 < 0$, where i_l is the modified spherical Bessel function [34].

The general basis regular in $r = 0$ for the solution space referred to the curvilinear basis $\mathbf{e}_r, \mathbf{X}_{lm}/|\mathbf{X}_{lm}|, \mathbf{e}_r \times \mathbf{X}_{lm}/|\mathbf{X}_{lm}|$, is then

$$\begin{bmatrix} -\frac{1}{Q_L^2} \frac{d}{dr} (j_l(Q_L r)) Y_{lm} \\ 0 \\ \frac{i\sqrt{l(l+1)}}{r Q_L^2} j_l(Q_L r) |\mathbf{X}_{lm}| \\ -\frac{4\pi\alpha}{\epsilon_\infty Q_L^2} J_l(Q_L r) Y_{lm} \end{bmatrix} , \begin{bmatrix} \frac{l(l+1)}{r Q_T^2} j_l(Q_T r) \\ 0 \\ -\frac{i\sqrt{l(l+1)}}{r Q_T^2} \frac{d}{dr} (j_l(Q_T r)) |\mathbf{X}_{lm}| \\ 0 \end{bmatrix} , \quad (7.23)$$

$$\begin{bmatrix} -\frac{\alpha l r^{l-1}}{\rho \omega_T^2 Q_T^2} Y_{lm} \\ 0 \\ \frac{i\sqrt{l(l+1)}\alpha}{\rho \beta_T^2 Q_T^2 r} r^l |\mathbf{X}_{lm}| \\ r^l Y_{lm} \end{bmatrix} , \begin{bmatrix} 0 \\ -\frac{i\sqrt{l(l+1)}}{Q_T} j_l(Q_T r) |\mathbf{X}_{lm}| \\ 0 \\ 0 \end{bmatrix} . \quad (7.24)$$

A similar basis, regular for $r \to \infty$, for the solution space can be written by substituting $r^l \to r^{-l-1}$ and $j_l \to h_l^{(1)}$.

To express the 'stress tensor' σ as a function of Ψ and Γ we add and subtract the term $\rho \beta_T^2 \nabla_j u_i / 2$ to Eq. (3.27) obtaining

$$\sigma_{ij} = -\rho(\beta_L^2 - 2\beta_T^2)\Psi g_{ij} - \rho \beta_T^2 \left[2\nabla_j u_i + \sum_k \epsilon_{ijk} \Gamma_k \right], \quad (7.25)$$

where g_{ij} is the metric tensor and ϵ_{ijk} the Levi-Civita tensor. The above equation has the same form in all orthogonal curvilinear coordinates. Thus, the components of $\sigma \cdot \mathbf{e}_r$ for a spherical quantum dot are given by

$$\sigma_{1r} = -2\beta_T^2 \rho \left[\frac{(\beta_L^2 - 2\beta_T^2)}{2\beta_T^2} j_l(Q_L r) - \frac{1}{Q_L^2} \frac{d^2}{dr^2}(j_l(Q_L r)) A_1 \right.$$
$$\left. + l(l+1) \frac{B_1}{Q_T^2} \frac{d}{dr}\left(\frac{j_l(Q_T r)}{r}\right) - \frac{\alpha C_1}{\rho \beta_T^2 Q_T^2} l(l-1) r^{l-2} \right] Y_{lm}, \quad (7.26)$$

$$\sigma_{2r} = \frac{i\rho \beta_T^2}{Q_T} D_1 \sqrt{l(l+1)} r \frac{d}{dr}\left(\frac{j_l(Q_T r)}{r}\right) |\mathbf{X}_{lm}|, \quad (7.27)$$

$$\sigma_{3r} = i2\rho \beta_T^2 \sqrt{l(l+1)} \left\{ -\frac{\alpha C_1}{\rho \beta_T^2 Q_T^2}(l-1) r^{l-2} - \frac{A_1}{Q_L^2} \frac{d}{dr}\left(\frac{J_l(Q_L r)}{r}\right) \right.$$
$$\left. + \frac{B_1}{Q_T^2}[j_l(Q_T r) - 2\frac{d}{dr}(\frac{1}{r}\frac{d}{dr}(r j_l(Q_T r)))] \right\} |\mathbf{X}_{lm}|, \quad (7.28)$$

where the indices (1, 2 or 3) represent the components along $\mathbf{e}_r$, $\mathbf{X}_{lm}$ and $\mathbf{e}_r \times \mathbf{X}_{lm}$, respectively.

Optical modes for the completely confined case

We now consider a quantum dot where the matching boundary conditions (3.47) at the interface can be applied, i.e. we use the mechanically rigid wall model for completely confined modes.

The conditions (3.44)-(3.46) are then:

$$\mathbf{u}|_{r=R} = 0, \quad (7.29)$$

$$\varphi|_{r=R-} = \varphi|_{r=R+}, \quad (7.30)$$

$$\epsilon_{a\infty}\frac{\partial\varphi}{\partial r}\Big|_{r=R-} = \epsilon_{b\infty}\frac{\partial\varphi}{\partial r}\Big|_{r=R+} . \tag{7.31}$$

These simplified boundary conditions have been discussed in Chapter 3, 4 and 5 and satisfactorily tested against microscopic lattice dynamical calculations.

As in the quantum well and quantum well wire cases here we have uncoupled and coupled oscillation modes.

Uncoupled modes

Using the space of solutions (7.23) and (7.24), the condition $\mathbf{u}(R) = \mathbf{0}$ leads to the eigenvalue equation

$$j_l(Q_T R = \mu_n) = 0 \quad ; \quad n = 1, 2, \ldots \tag{7.32}$$

corresponding to transverse modes with frequencies given by

$$\omega^2 = \omega_T^2 - \beta_T^2 \left(\frac{\mu_n^{(l)}}{R}\right)^2 . \tag{7.33}$$

These solutions are modes vibrating along the $\mathbf{X}_{lm}$ direction and they have no electrostatic potential associated with them. The solutions are completely decoupled from the other curvilinear components $\mathbf{e}_r$ and $\mathbf{e}_r \times \mathbf{X}_{lm}$ and have a purely transverse character with respect to the sphere radius. That is, they are purely tangential modes. The mechanical amplitude is given by

$$\mathbf{u} = -iD_1\sqrt{l(l+1)}\frac{R}{\mu_n^{(l)}} j_l\left(\frac{\mu_n^{(l)}}{R}r\right) \mathbf{X}_{lm}(\theta, \varphi) . \tag{7.34}$$

Coupled modes

The matching conditions (7.29)–(7.31) applied to the other two components of the vector $\mathbf{u}$ and to the electrostatic potential $\varphi(\mathbf{r})$ yield a homogeneous system of linear equations for the constants A_1, B_1, C_1 and C_2. For these modes the following secular equation is found:

$$v j_l'(v) F_l(\mu) = l(l+1) j_l(v) G_l(\mu) , \tag{7.35}$$

where

$$F_l(\mu) = \frac{\omega_L^2 - \omega_T^2}{\beta_T^2}\left(\frac{R}{\mu}\right)^2 l[\mu j_l'(\mu) - l j_l(\mu)] + \left(\frac{\epsilon_{a\infty}}{\epsilon_{b\infty}}(l+1) + l\right)[\mu j_l'(\mu) + j_l(\mu)] ,$$

$$\tag{7.36}$$

$$G_l(\mu) = \frac{\omega_L^2 - \omega_T^2}{\beta_T^2} \left(\frac{R}{\mu}\right) \frac{\epsilon_{a\infty}}{\epsilon_{b\infty}} [l j_l(\mu) - \mu j_l'(\mu)] + \left(1 + \frac{\epsilon_{a\infty}}{\epsilon_{b\infty}}(l+1)\right) j_l(\mu) , \tag{7.37}$$

with

$$\mu = Q_T R , \quad \nu = Q_L R .$$

Then the vector **u** has two components: one along $\mathbf{e}_r$ and the other one along $\mathbf{e}_r \times \mathbf{X}_{lm}$, that is

$$\mathbf{u} = u_1(r) Y_{lm}(\theta, \varphi) \mathbf{e}_r + u_3(r) \mathbf{e}_r \times \mathbf{X}_{lm}(\theta, \varphi) , \tag{7.38}$$

with

$$u_1 = A \left[-\frac{d}{dr}(j_l(Q_L r)) + \frac{l+1}{r} p_l j_l(Q_T r) - l \frac{t_l}{R} \left(\frac{r}{R}\right)^{l-1} \right] , \tag{7.39}$$

$$u_3 = -iA \frac{\sqrt{l(l+1)}}{r} \left[-j_l(Q_L r) + \frac{p_l}{l} \frac{d}{dr}(r j_l(Q_T r)) - t_l \left(\frac{r}{R}\right)^l \right] , \tag{7.40}$$

$$p_l = \frac{\nu j_l'(\nu) - l j_l(\nu)}{l j_l'(\nu) - \mu j_l(\nu)} , \tag{7.41}$$

$$t_l = \frac{\gamma_0}{\mu^2} \frac{[\nu j_l'(\nu) + (l+1) \frac{\epsilon_{b\infty}}{\epsilon_{a\infty}} j_l(\nu)]}{1 + \frac{\epsilon_{b\infty}}{\epsilon_{a\infty}}(l+1)} , \tag{7.42}$$

$$\gamma_0 = \frac{\omega_L^2 - \omega_T^2}{\beta_T^2} R^2 ,$$

and A is a normalisation constant.

The most important contribution to one-phonon Raman scattering in spherical quantum dots corresponds to $l = 0$ [37]. In this case the $\mathbf{X}_{00}$ component is absent and thus only radial modes are obtained. The secular equation for the eigenvalues is then

$$\tan(\nu_n) = \nu_n , \quad n = 1, 2, \ldots \tag{7.43}$$

and the corresponding vibrational amplitude is

$$\mathbf{u}(r) = A j_1\left(\nu_n \frac{r}{R}\right) \mathbf{e}_r . \tag{7.44}$$

The modes have then a purely longitudinal character and the eigenfrequencies are

$$\omega^2 = \omega_L^2 - \beta_L^2 \left(\frac{\nu_n}{R}\right)^2. \qquad (7.45)$$

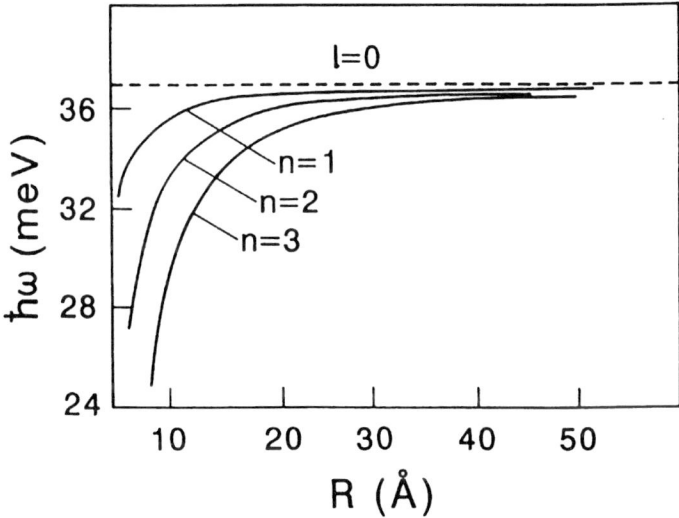

Figure 7.1 The optical mode energies of the first three $l = 0$ solutions of Eq. (7.43) as a function of R. The parameters of GaAs have been used in the calculation.

In the limiting case of R large Eq. (7.35) is reduced to

$$j'_l(Q_L R) j'_l(Q_T R) \left[\frac{\omega_L^2 - \omega_T^2}{\beta_T^2} \frac{l}{Q_T} + Q_T \left(1 + \frac{\epsilon_{2\infty}}{\epsilon_{1\infty}}(l+1)\right)\right] = 0. \quad (7.46)$$

It follows from the above equation that [32] for $\mu, \nu \gg 1$

$$\omega^2 = \omega_T^2 \frac{\epsilon_{a0} l + \epsilon_{b\infty}(l+1)}{\epsilon_{a\infty} l + \epsilon_{b\infty}(l+1)} \qquad (7.47)$$

According to Eq. (7.47) a series of modes with frequencies between ω_T and ω_L is obtained. For $l = 1$, (7.47) gives

$$\omega_F^2 = \omega_T^2 \frac{\epsilon_{a0} + 2\epsilon_{b\infty}}{\epsilon_{a\infty} + 2\epsilon_{b\infty}}. \tag{7.48}$$

This represents the so-called Fröhlich frequency which corresponds to a uniform polarisation of the sphere. The modes $l > 1$ are called surface modes. In the case of $R \gg Q_T^{-1}$ and $R \gg Q_L^{-1}$ the mechanical matching boundary conditions can be neglected and the modes (7.47) are basically the same as obtained by imposing only electrostatic boundary conditions on φ. If $R \sim Q_T^{-1}$ or Q_L^{-1} then the effect of the mechanical matching conditions becomes important and Eq. (7.47) is no longer valid.

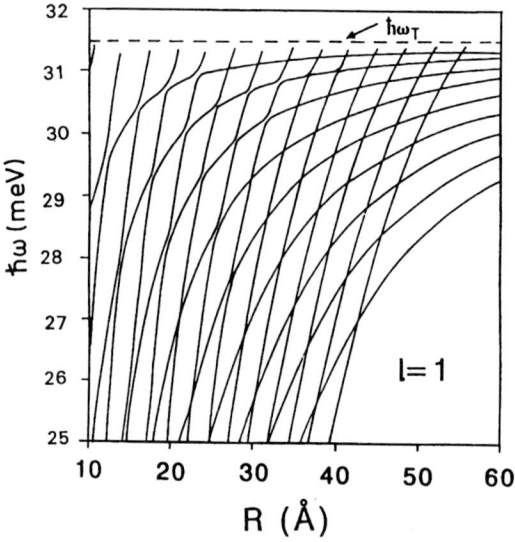

Figure 7.2 Optical vibrational energies as a function of radius R in the frequency range $\omega < \omega_T$ for $l = 1$. The numerical calculations were performed for GaAs embedded in AlAs.

Figure 7.1 shows the optical vibrational energies of the first three modes of a GaAs dot in AlAs for $l = 0$ as a function of R. The $n = 1$ mode occurs basically at the $\hbar\omega_L$ frequency of the dot material, except for $R \leq 20$ Å. The modes of higher n begin to decrease in frequency at larger radii.

Figure (7.2) shows the mode energies ($\hbar\omega < \hbar\omega_T$) as a function of radius R for $l = 1$ of a GaAs dot. The strong mixing between the u_1 and u_3 components is quite apparent. It is clear that whenever a curve resembling a longitudinal mode dispersion curve, where the vibrational amplitude is a vector $\mathbf{u}$ along $\mathbf{e}_r$, approaches the corresponding transverse-mode curve an anticrossing takes

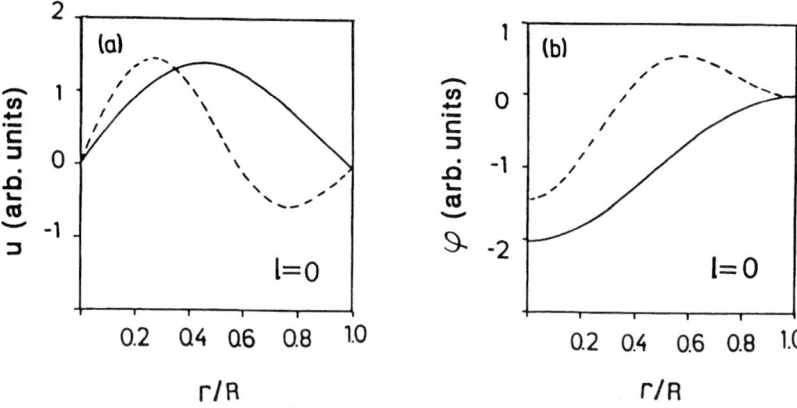

Figure 7.3 The vibrational amplitude u (a) and the electrostatic potential φ (b) for $l = 0$ in a GaAs quantum dot as a function of r for $R = 20$ Å. The values of $\hbar\omega = 36.4$ meV (solid line) and $\hbar\omega = 35.8$ meV (dashed line) were used in the calculation.

place and the behaviour changes from longitudinal to transverse and viceversa.

Figure (7.3) illustrates the electrostatic potential φ and the vibrational amplitude u for $l = 0$ as a function of r for $R = 20$ Å and the first two modes ($n = 1$ and 2) in a spherical GaAs quantum dot.

The components $u_1(r)$ and $u_3(r)$ and φ for the same case as Figure 7.3 are shown in Figure 7.4 for $l = 1$ and the first two mode energies $\hbar\omega = 36.3$ meV and $\hbar\omega = 35.4$ meV, respectively.

We now discuss the optical phonon dispersion relations for the $l = 1$ and 2 modes of a CdS spherical microcrystallite in glass. The $u = 0$ matching condition implies physically that the Cd atoms at the surface of the sphere are bonded to the O atoms of the glass [37]. For bulk CdS the dispersion of the transverse modes is predicted theoretically to be positive [38]. Hence the transverse confined modes have frequencies larger then the bulk transverse zone centre frequency. Figure 7.5 illustrates the dependence of the $l = 1$ normal frequencies on the radius. The parameters of [37] were used for the calculation. Between ω_T and 255 cm^{-1} we can observe anticrossing of modes due to confined transverse components which tend to ω_T at large radii. Above $\omega = 255$ cm^{-1} one can see the contributions due to the confined longitudinal modes with some bending in the dispersion at 264 cm^{-1} caused by the interaction with the contribution of the electrostatic surface mode. This surface mode corresponds to the Fröhlich modes [9]. At a frequency of 255 cm^{-1} the transverse wavevector component is at the L-point of the Brillouin Zone in bulk CdS. Therefore, the wavevector component for the transverse modes must be continued along

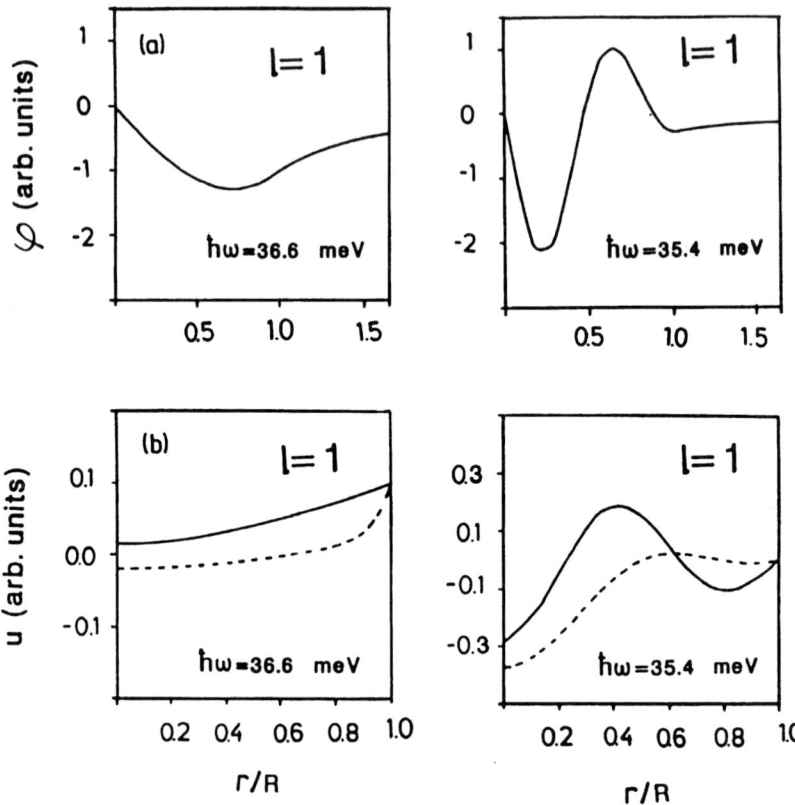

Figure 7.4 The same as Figure 7.3 for $l = 1$ and mode energies $\hbar\omega = 36.6$ meV and $\hbar\omega = 35.4$ meV. The amplitude $u_1(r)$ is represented by full lines while $u_3(r)$ is given by dashed lines.

the complex Brillouin Zone to obtain frequencies higher than 255 cm^{-1}. The complex wavevector damps out the contribution of the transverse modes to the dispersion relation above 255 cm^{-1}.

Figure 7.6 shows the $l = 2$ mode frequencies, similar to those in Figure 7.5, also as a function of the quantum dot radius. The effect of the surface mode is at a frequency slightly above 269 cm^{-1}.

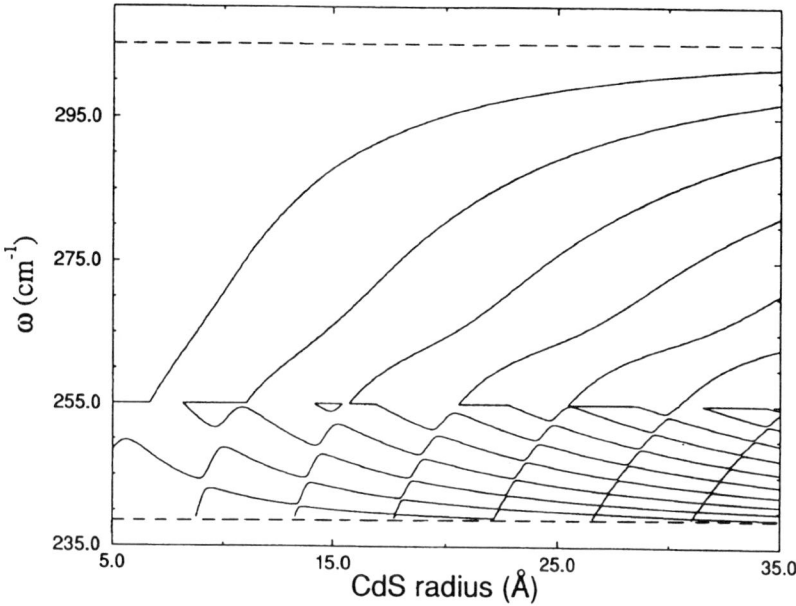

Figure 7.5 The radial dependence of the $l = 1$ vibrational modes frequencies for a CdS quantum dot embedded in glass. The longitudinal optical and transverse optical bulk frequencies are indicated by dashed lines. See discussion in text.

Free surface condition

Self-organisation effects on semiconductor surfaces are now considered a way for in-situ creation of quantum dots [39]. Uniform arrays of quantum dots which can be considered isolated and coherent, have been reported in III-V heterostructures. Quantum dots of InSb, InAs, InGaAs, etc. are grown on different surface substrates [22]. These can be considered as systems with free surface condition and in a first approximation we will treat them as quantum dots having spherical geometry.

The conditions (3) after tedious but straightforward manipulations leads to the following results

Decoupled modes with the dispersion relation

$$\omega^2 = \omega_T^2 - \beta_T^2 \left(\frac{\mu_n^{(l)}}{R} \right)^2, \tag{7.49}$$

where the $\mu_n^{(l)}$ are determined by the condition

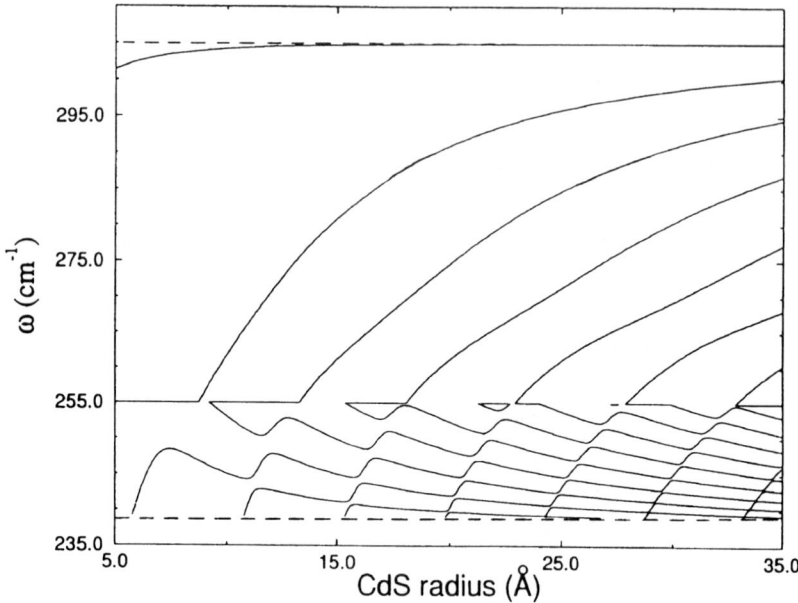

Figure 7.6 The same as in Figure 7.5 for $l = 2$.

$$j_l(\mu) = \mu j_l'(\mu) \ . \qquad (7.50)$$

These modes have purely transverse character with respect to the radius vector $\mathbf{e}_r$ on the sphere and the vector $\mathbf{u}$ is along $\mathbf{X}_{lm}$ direction. The mechanical vibration presents the same behaviour as that given by Eq. (7.34).

For the *coupled modes* the following equation for the eigenmodes is obtained:

$$a_l\, j_l(Q_L R)\, j_l(Q_T R) + b_l Q_T R\, j_l(Q_L R)\, j_l'(Q_T R) + c_l Q_L R\, j_l'(Q_L R)\, j_l(Q_T R)$$
$$= d_l R^2 Q_L Q_T\, j_l'(Q_L R)\, j_l'(Q_T R) \ , \qquad (7.51)$$

where

$$\begin{aligned} a_l \;=\; & l(l^2 - 1)(l+2)[\bar{\epsilon}(\gamma_0 l + l + 1) + l] \\ & + \frac{Q_T^2 R^2}{2}[\frac{l}{2}(\gamma_0 + 1)^2 Q_T^2 R^2 + \bar{\epsilon}(l+1)(\gamma_0 + 1)\left(\frac{Q_T^2 R^2}{2} + 1\right) \\ & - \bar{\epsilon} 2l(l+1)(\gamma l + l + 1) \\ & - l(2l^2 + 2l - 1 + 2\gamma_0 l^2 + 3\gamma_0 l + \gamma_0^2 l - \gamma_0)] \ , \qquad (7.52) \end{aligned}$$

$$b_l = \frac{\gamma_0 + 1}{2} Q_T^2 R^2 [l(\gamma_0 + 1) + (l+1)\bar{\epsilon}] - \bar{\epsilon}\gamma_0 \gamma l(l^2 - 1)(l+2) , \quad (7.53)$$

$$c_l = Q_T^2 R^2 [(\gamma_0+1)l+(l+1)\bar{\epsilon}]+(l-1)(l+2)[l(\gamma_0 l-1)-(l+1)\bar{\epsilon}] , \quad (7.54)$$

$$d_l = (l-1)(l+2)[l(\gamma_0 + 1) + \bar{\epsilon}(l+1)] \quad (7.55)$$

and $\bar{\epsilon} = \epsilon_{b\infty}/\epsilon_{a\infty}$

Fröhlich type Hamiltonian

The set of eigenfunctions $\mathbf{u}_{nlm}(\mathbf{r})$ given by (7.38) can be normalised so that

$$\sum \mathbf{u}_{n'l'm'}^* \cdot \mathbf{u}_{nlm} \, \rho \, d^3r = \delta_{nn'}\delta_{mm'}\delta_{ll'} . \quad (7.56)$$

The electrostatic potential φ_{nlm} derived from Eq. (7.3) leads directly to the Fröhlich-type electron-phonon Hamiltonian. Following the procedure described in Chapter 3 and for the case of complete confinement the interaction Hamiltonian is given by the equation.

$$\hat{H}_F = e\hat{\phi}_F$$
$$= \sum_n \sum_{l=0}^{\infty} \sum_{m=-l}^{l} \frac{C_F}{\sqrt{R}} \varphi_{nl}(r) Y_{lm}(\theta,\varphi) \left(\hat{b}_{nlm} + \hat{b}_{nlm}^\dagger\right) \quad (7.57)$$

with

$$\varphi_{nl} = \frac{\sqrt{R}}{|u|} \begin{cases} j_l(\nu_n \frac{r}{R}) - \frac{\nu_n j_l'(\nu_n)+(l+1)\frac{\epsilon_{b\infty}}{\epsilon_{a\infty}} j_l(\nu_n)}{1+\frac{\epsilon_{b\infty}}{\epsilon_{a\infty}}(l+1)} (\frac{r}{R})^l & ; \quad r < R \\ \\ -\frac{\nu_n j_l'(\nu_n)-l j_l(\nu_n)}{1+\frac{\epsilon_{b\infty}}{\epsilon_{a\infty}}(l+1)} (\frac{R}{r})^{l+1} & ; \quad r > R \end{cases}, \quad (7.58)$$

and the term $|u|$ represents the norm of the vibrational amplitude, given by

$$|u|^2 = \int_0^R \left\{ \left[-\frac{d}{dr} j_l\left(\nu \frac{r}{R}\right) + \frac{l+1}{r} p_l j_l\left(\frac{\nu r}{R}\right) - \frac{l t_l}{R}\left(\frac{r}{R}\right)^{l-1} \right]^2 \right.$$
$$\left. + \frac{l(l+1)}{r^2} \left[-j_l\left(\nu \frac{r}{R}\right) + \frac{p_l}{l} \frac{d}{dr}\left(r j_l\left(\frac{\nu r}{R}\right)\right) - t_l\left(\frac{r}{R}\right)^l \right]^2 \right\} r^2 dr, l \neq 0$$

$$(7.59)$$

The important case $l = 0$ is reduced to:

$$\varphi_{n0} = \frac{\sqrt{R}}{\nu_n} \left[j_0\left(\nu_n \frac{r}{R}\right) - j_0(\nu_n) \right] \, , r < R \qquad (7.60)$$

and zero for $r > R$.

For the coupled solutions we find again — this time with spherical symmetry — the usual pattern of modes having both, confined and free surface character, the details changing as the sphere radius changes.

In Eq. (7.57) $\hat{b}_p$ and $\hat{b}_p^\dagger$ are annihilation and creation operators for vibrational modes in the state p. The Fröhlich polar electron-phonon interaction has been calculated in [7, 11]. In Ref [11] the classical dielectric model, neglecting dispersion of the bulk phonon branches, has been used to derive an independent set of longitudinal vibrational and surface modes. These have been evaluated by applying electrostatic boundary conditions using the formalism of Ref [39]. On the other hand by applying the hydrodynamical model [8] a Hamiltonian for the electron confined longitudinal modes interaction has been obtained upon application of spherical boundary conditions. In both these cases the coupling between longitudinal and transverse modes was not taken into account and they were unable to describe the mixing of the confined and surface modes.

A clear experimental confirmation concerning the influence of electrostatic and mechanical boundary conditions on the vibrational modes of semiconductor nanocrystals can be found elsewhere [16]. The Raman scattering and far infrared absorption spectra were measured for PbS nanocrystals by observing the coupled and mixed vibrational modes in correspondence with the theoretical model which accounts for the mechanical and electrostatic matching conditions at nanocrystal interfaces. Thus this model is seen to provide a satisfactory account of all the key features of polar optical modes for all the archetypal geometries.

References

1. C.W.J. Beenakker and H. van Houten, in Solid State Physics: Semiconductor Heterostructures and Nanostructures, edited by H. Ehrenreich and D. Turnbull, Academic Press, San Diego (1991), V **44**.

2. Y. Wang and N. Herron, J. Phys. Chem. **95**, 525 (1991).

3. Y.S. Tang and C.M. Sotomayor Torres, R.A. Kubiak, T.E. Whall, E.H.C. Parker, H. Presting and H. Kibbel, J. of Electronic Materials **24**, 99 (1995).

4. H. Yükselici and P.D. Persans and T.M. Hayer, Phys. Rev. **B52**, 11763 (1995).

5. E. Duval, A. Boukenter, and B. Champagnon, Phys. Rev. Lett. **56**, 2052 (1986).

6. J.J. Shiang, A.N. Goldstein and A.P. Alivisatos, J. Chem. Phys. **92**, 3232 (1990).

7. A.I. Ekimov, Physica Scripta **T39**, 217 (1991).

8. S. Nomura and T. Kobayashi, Phys. Rev. **B45**, 1305 (1992).

9. S. Nomura and T. Kobayashi, Solid State Commun. **82**, 335 (1992).

10. D. J. Norris, A. Sacra, C.B. Murray, and M.G. Bawendi, Phys. Rev. Lett. **72**, 2612 (1994).

11. M.C. Klein, F. Hache, D. Ricard and C. Plytzanis, Phys. Rev. **B42**, 11123 (1990).

12. B. Champagnon, B. Andrianasolo, and E. Duval, Materials Science and Engineering **B9**, 417 (1991).

13. B. Champagnon, B. Andrianasolo, and E. Duval, J. Chem. Phys. **94**, 5237 (1991).

14. A.V. Baranov, Ya. S. Bobovich and V.I. Petrov. Solid State Commun. **83**, 957 (1992).

15. A. Mlayah, A.M. Brugman, R. Carles, J.B. Renucci, M. Ya. Valakh and A.V. Pogorelov, Solid State Commun. **90**, 567 (1994).

16. T.D. Krauss, F.W. Wise and D.B. Tanner, Phys. Rev. Lett. **76**, 1376 (1996).

17. N. F. Borelli and D.W. Smith, J. Non-Cryst. Solids **180**, 25 (1994).

18. W. Hansen, T.P. Smith, K.Y. Lee, J.A. Brumm, C.M. Kruedler, J.M. Hong and D.P. Kern, Phys. Rev. Lett. **62**, 2168 (1989).

19. Al. L. Efros, A.I. Ekimov, F. Kozlowski, V. Petrova-Koch, H. Schmidbaur and S. Shumilov, Solid State Commun. **78**, 853 (1991).

20. S. Hayashi and H. Kanamori, Phys. Rev. **B26**, 7079 (1982).

21. Ch. Sikorski and U. Merkt, Phys. Rev. Lett. **62**, 2164 (1989).

22. B.R. Bennett, B.V. Shanabrook and R. Magno, Appl. Phys. Lett. **68**, 958 (1996).

23. B.R. Bennett, R. Magno and B.V. Shanabrook, Appl. Phys. Lett. **68**, 505 (1995).

24. T. Okadad, T. Iwaki, K. Yamamoto, H. Kasahara and K. Abe, Solid State Commun. **49**, 809 (1984).

25. S. Hayashi, M. Ito and H. Kanamori, Solid State Commun. **44**, 75 (1982).

26. L.H. Campbell and P.M. Fauchet, Solid State Commun. **58**, 739 (1986).

27. Y. Sasaki and C. Horie, Phys. Rev. **B47**, 3811 (1993).

28. C.J. Sandroff and L.A. Farrow, Chem. Phys. Lett. **130**, 458 (1986).

29. Y. Sasaki, Y. Nishina, M. Sato and K. Okamura, Phys. Rev. **B40**, 1762 (1989).

30. H. Fröhlich, *Theory of Dielectrics*, Oxford University Press, Oxford (1948).

31. R. Fuchs and K.L. Kliewer, Phys. Rev. **140**, A2076 (1970).

32. R. Ruppin and R. Englman, Rep. Prog. Phys. **33**, 149 (1970).

33. M. Babiker, J. Phys. C **19**, 683 (1986).

34. M. Abramovitz and I.A. Stegun, *Handbook of Mathematical Functions*, Dover, New York (1972).

35. P.M. Morse and H. Feshbash, *Methods of Theoretical Physics*, Mc Graw Hill, New York (1953).

36. J.D. Jackson, *Classical Electrodynamics*, Wiley, New York (1975).

37. M.P. Chamberlain, C. Trallero-Giner and M. Cardona, Phys. Rev. **B51**, 1680 (1995).

38. M.A. Nusimovici and J.L. Birman, Phys. Rev. **156**, 925 (1967).

39. J. Tersoff and R.M. Tromp, Phys. Rev. Lett. **70**, 2782 (1993).

Index

$(CdS)_x(CdSe)_{1-x}$, 145
$Al_xGa_{1-x}As$, 8, 93
$Al_xGa_{1-x}As$, 85
AlAs, 8, 34, 91
alloys, 10
AlSb, 145
annihilation operator, 2
atomic displacements, 92

Bloch, 7
Bloch condition, 100
Born von Karman, 5, 7
Born-Huang, 26
Born-Huang equations, 14
bulk, 1

CdS, 145
CdSe, 145
CdSSe, 145
CdZnSe, 145
collective coordinates, 4
commutation relation, 18
commutation relations, 45
commutation rule, 135
completeness, 36
couple modes, 117
coupled mode, 119
creation operator, 2
crystal anisotropy, 41
cutoff, 17

dielectric continuum model, 123
dispersion relation, 128, 129
double barrier structures, 106
double heterostructure, 88

eigenfunctions, 131

electron-phonon Hamiltonian, 139
electron-phonon interaction Hamiltonian, 17, 44, 135
envelope-function theory, 14
Esaki, 22
Euler-Lagrange, 25
even potential modes, 87

Fröhlich frequency, 145, 154
Fröhlich-type electron phonon Hamiltonian, 159
free oscillations, 34
free standing wire, 136
free surface condition, 157
free-standing wire, 123

GaAs, 8, 34, 85, 93, 120, 124, 145
GaP, 145
GaSb, 23, 120, 145
Ge, 23, 145
Green function, 56, 59

Hamiltonian density, 26
Hankel spherical functions, 146
heavy interface, 115
hermiticity, 36
high-frequency dielectric tensor, 27
hydrodynamic model, 123

InAs, 23, 120, 145
InGaAs, 145
InP, 145
InSb, 145

Lagrangian, 3, 6
Lagrangian density, 24
light interface, 115
linear chain, 1, 2

long wave limit, 13

matching boundary conditions, 116
matching conditions, 32, 76
matching formula, 54, 55, 64, 66, 67, 80
molecular-beam epitaxy, 123
multilayer structure, 66

odd potential modes, 86
orthogonality, 36

PbI_2, 145
PbS, 145
phonon-assisted tunnelling, 108
piezoelectric waves, 16
planar vibrational modes, 114
Poisson equation, 16
polarisation, 14, 29
potentials, 83
projector, 63, 65, 66, 68, 70, 80
purely longitudinal modes, 88, 127
purely transverse mode, 119
purely transverse modes, 88, 128
purely transverse solutions, 84

quantum dots, 145, 146, 149, 150, 152, 157
quantum well, 85
quantum well wire, 123

quantum well wires, 123
quasi-longitudinal modes, 96
quasi-transverse modes, 96

scalar function Ψ, 30
secular equation, 138, 152
selective ion implantation, 123
SGFM, 52
shear horizontal wave, 96
Si, 23, 145
SiC, 145
SiGe, 145
spatial dispersion, 56
speed of sound, 13
spherical harmonics, 146
static dielectric tensor, 27
stiffness coefficients, 56
strained superlattices, 145
stress tensor, 28
superlattices, 98
surface modes, 154

t-matrix, 61, 62
transverse horizontal, 56
transverse horizontal mode, 116
Tsu, 22
two-mode model, 9

vector function Γ, 30